SpringerBriefs in Statistics

More information about this series at http://www.springer.com/series/8921

Jordi Vallverdú

Bayesians Versus Frequentists

A Philosophical Debate on Statistical Reasoning

Jordi Vallverdú
Faculty of Philosophy and Arts
Universitat Autònoma de Barcelona
Bellaterra (Cerdanyola del Vallès)
Spain

ISSN 2191-544X   ISSN 2191-5458 (electronic)
SpringerBriefs in Statistics
ISBN 978-3-662-48636-8   ISBN 978-3-662-48638-2 (eBook)
DOI 10.1007/978-3-662-48638-2

Library of Congress Control Number: 2015954586

Mathematics Subject Classification (2010): 00A30, 01Axx, 62-03

Springer Heidelberg New York Dordrecht London

Printed on acid-free paper

Springer-Verlag GmbH Berlin Heidelberg is part of Springer Science+Business Media
(www.springer.com)

To my amazing son Cai,
I love you with all my heart and mind
with no chance of any other true feeling
towards you.

Preface

Despite possible misunderstanding which I assure readers I have not created, the reader might think, from its title that this book is a piece of research about the Philosophy of Statistics, a new brick into the great "Chinese wall" of the unending philosophical debates about statistics. But this is not the case. Taking into account that today philosophers still do not agree on the basic notions of "mind," "cause," "evidence," "experience," or "truth," it would be a waste of time to devote our interests to the fuzzy and even impolite debates about *what* things *are*. Nevertheless, we need to talk, solve, and think about several domains that affect our lives. For example, why do things happen? Our minds are just used to accepting any possible result: Our brains predict possible outcomes of several sensorimotor actions at the same time our minds are trying to discover some kind of order in our world (internal, external, cultural, real, or invented, does not matter). For all these reasons, this book tries to deal with several connected questions: How do our minds operate with quantities? What are the most natural ways to deal with information? How can several epistemological models efficiently satisfy the numerical analysis of reality? Why are certain events only feasibly understandable through the analysis of big numbers? Some of these questions led us to the naturalization of statistics, a project to which I have contributed scarcely but with high intensity in this book. We will also learn how classic philosophical debates, like those related to causality or determinism, have been updated and continued by modern statistical thinkers or practitioners. The debate about the best and parsimonious methods is still present. During the twenty-first century, we are ceasing to do things with words and we are starting to explain concepts more and more with numbers. The last remnants of qualitative analysis are being colonized by numerical methods. Thus, sociologists, psychologists, philosophers, and archeologists, among other field experts, are entering into a new computational, statistical, and experimental era. For all the previous reasons, this cannot be understood as a book on Philosophy or History of Statistics, although part of the purposes of this book does belong in these fields. The overspecialization leads very frequently to experts unable to see the forest for the trees, and then, their scholarship becomes an academical product, empty scholasticism. When we discuss about the numerical approach to the reality. we are not acting as mystical Neopythagoreans; instead, we are analyzing how the numerization process is created and how the rules work with these numbers. We are talking about a natural property of the cognitive systems (to deal with quantities) that has been heavily improved with symbolic tools. After centuries of intense scientific revolution, it is lamentable that rather than innovating we are following the same paths that made our ancestors venture their lives into the African savannah: to understand the world and survive adding "best" or more accurate meanings to our lives. The greatness and, at the same time, the main defect of this book are to explain briefly that while we are entering into a new paradigm of (numerical) research at the same time, we are not leading the need to solve our practical/conceptual necessities. Big data, deep mining, big mechanisms, e-Science, or computational simulations are only possible thanks to a new era of hardware and mindware: Statistics are the backbone of this revolution. Numbers now trace the path of the arrow of human evolution. Let me introduce you to some facts, debates, ideas, and also doubts about how we can understand our world. This is the true meaning of this book.

Arenys de Munt, Catalonia   Jordi Vallverdú
July 2015

Acknowledgments

My gratitude to Prof. Miodrag Lovric for believing in my work and accepting to publish the initial documents led to this small piece of conceptual revision on the theory of statistics. I must extend this gratitude to Veronika Rosteck, editor from Springer editorial, for her *huge and close to infinite* patience with me and my complex personal circumstances all along these last years. Three anonymous reviewers did also their work trying to improve my ideas.

I also need to thank the JSPS organization for a splendid fellowship to make research at Kyoto University (October–November 2011). I employed part of my time thinking on the statistical roots of our research as well as enjoying one of the most incredible cultures of our world, the Japanese, close to nature as well as to robotic and computational realities. Without this research funding and the pleasant moments spent thinking early in the morning by the shore of Kamogawa River, this work could not be possible.

I thank Profs. Stephen M. Stigler and Gustavo Deco for their time and their kind answers to my questions as well as for their suggestions; to Stefan Conrady (from Bayesia) for his answers and very specific bibliographical assistance; to Seamus Bradley and Carl Hoeffer for their fast answer to my arisen questions or demands on intellectual help; to Anja Stemme for her Old German assistance as well as for our very exciting online intellectual debates; to Prof. Judea Pearl for his theoretical suggestions and ideas; and very, very especially to Prof. David Gillies for his nice offering to discuss and share our academic interests as well as his full draft reading as well as numerous corrections and suggestions. Even without agreeing with me about several aspects of the philosophical debate on statistics, Prof. Gillies was extremely generous offering his precise, deep, and politely expressed ideas contributing definitively to the improvement of the content of this book. I feel somehow guilty stealing research time from great researchers like them. Finally, I thank Huma Shah for her help at the end of the revision process. Without her, this book would not reach your hands or screens (Hi, ciberpirates).

I thank the research group Tecnocog, headed by Prof. Anna Estany, and his research project "Innovación en la práctica científica: enfoques cognitivos y sus consecuencias filosóficas" (FF2011-23238)," funded by MCI. I also must give my gratitude to 2014-SGR-412 research group on Epidemiology and Philosophy, to which I belong, for sharing with me their ideas and to discuss constructively my ideas on causation and statistical reasoning in epidemiological contexts.

I thank my mom, Carme, for a whole life devoted to the daily existence. I have still to learn so much about you!

I thank my sons, Sujan and Cai. I really cannot be more proud of you, boys. Perhaps you will never read any of my books, but even in that (im) probable situation I still will love you with all my heart... ;-) Life shows you clearly what is important and what is not. And you are the backbone of my life. I have no doubt about this.

I thank Sarah, my wife, for allowing me to have a new opportunity to find happiness as well as for her encouragement and warm love all throughout a very complex moment of my life. I will not promise anything to you, because words are only words, but I declare you that I want share with you my whole present. Sometimes, less is more.

Finally, I thank my (social) brain, for letting me to be inside such fascinating ideas about the reality of things, despite of its reluctance to be permanently at the limits of the thinkable. It is a marvelous machine that escapes to my analysis but offers me magic moments of knowledge, astonishment, and pleasure. It is the greatness of evolution and emergence of life. Hazard, matter, and contextual interactions.

Contents

Chapter 1
Some Questions to Begin with

Abstract First chapter analyzes how living systems such as amoebae, insects, fishes, chicks, or dolphins are able to deal with numbers without any symbolic system, explaining the notions of "subitization" or "numerosity," among others. At the same time, the cognitive limits for humans in relation to number processing are explored, especially those expressed by kids. Once provided this basic naturalistic framework for minds and numbers, the concepts of ignorance, chance, and statistics are introduced as well as their related basic philosophical debates.

Keywords Subitization · Numerosity · Animal cognition · Cognition · Natural statistics · Ignorance · Chance · Determinism

A book about statistics is, at the end, a book about numbers and how we establish relationships between these numbers and some events in the world. Perhaps, it is a problem of my background as philosopher, but I think that before we study the schools of statistics with more or less detail, we should reconsider what we know about numbers, minds, and chance.

1.1 Numbers and/or Neurons

Our knowledge comes from two different sources:

On the one hand, from experiential data (from own observations, experiments, prerecorded first- or second-hand data stored in books/computers/oral tales,…) and on the other, from mental mechanisms to obtain connections, regularities, patterns, or relationships among data. It is a dynamic feedback process in which the emphasis into one or another side of both prominent sources marks the classification as realist/idealist or similar categories. In some cases, this knowledge is expressed linguistically, but in some special cases by numbers. And the laws and semantics of numbers are not the same as that of words.

From our common sense and natural understanding, we can infer that human minds are not naturally designed to deal with numbers and let me explain why.

The maximum amount of elements we can visually identify as an exact "number" is 4. As I have written elsewhere (Vallverdú 2010), the human being does not have an immediate perception of numbers greater to 4 (Ifrah 1999, Chap. 1). That is to say, that he or she has a direct clear knowledge from 1 to 4. We can even note that diverse native groups (not polluted by annoying Western intruders) of the Australian Continent, Asia, America, and Africa only have in their respective languages terms like "one," "two," and "many." In several areas of our planet,[1] survival does not require (nor has it required) more than a short range of numbers from 1 to 3 or 4. This range is sufficient for necessary precision of basic activities. Moreover, first attempts at counting are deeply influenced by bodily structure: across cultures, primitive counting on fingers and toes led to the establishing of base 5 (following the number of fingers of a hand), base 10 (two hands), or base 20 (two hands plus two toes) (Gvozdanovic 2006: 736). Counting by hand is also present in the meaning of Proto-Indo-European words for "5" and "10." Consequently, at a certain basic level we can affirm that numbers are the result of our body structure and are physically embedded into it, which is a tool to shape gradually higher cognition from our senses. Consequently, sensorimotor experiences and abstract mathematical concepts are connected, and maths was initially embodied (Bender and Beller 2012).[2] In a nutshell, numbers are human.[3]

Beyond the body constraints at the beginning of counting tools (in this case), there are several pending conceptual aspects about numbers and counting:

(a) Ostensive definition and mimics: It generally expressed as "defining by finger pointing," an ostensive definition implies the definition or identity of a concept through the finger pointing at the object itself. According to Wittgenstein (1953, §6):

> I will call it "ostensive teaching of words". —I say that it will form an important part of the training, because it is so with human beings; not because it could not be imagined otherwise.) This ostensive teaching of words can be said to establish an association between the word and the thing.

---

[1] Surprisingly, there is at least one human community that count in the language *without numbers*, even without simple words like "one" or "two": the Hi'aiti'ihi (called "Pirahã" by experts), an indigenous hunter-gatherer tribe of Amazon natives, a subgroup of the Mura, who mainly live on the banks of the Maici River in Brazil's Amazonas state. Although they perform several complex activities (tool making, food location, and qualities, social shared activities, etc., their language— one of the most simple phonetically—has no cardinal or ordinal numbers (only approximate descriptions or amounts, like "few," "some," and "more"). Dan Everett, after several years of field research, wrote an impressive and seminal 200-page chapter on Pirahã language published in Derbyshire and Pullum (1986). See also Frank et al. (2008), and finally, I recommend Gordon (2004), to fulfill the debate.

[2] About embodiment and cognition, I recommend the first section of Carsetti (2010) to find a naturalistic and evolutionary view about how cognition is bodily constrained.

[3] I accept that someone could ask me: Can numbers exist in the non-human world? My answer is "yes, of course," but numerosity is directly related to ontological views on reality such as entities' stability, mass or volume conservation, among other possible aspects. So, numbers as we understand them can only exist in our cultural minds.

In the identity between one mark into a bone and one day or between one object and one finger, it is a highly symbolic process, something close to a friendly extension of imitation. We learn looking at the place where the others eyes gaze or fingers/hands/arms point to. We know then the intentions and interests of the others toward one external thing to our minds. This thing is also explained by the others with more bogy gestures, voice intonation or even language. Then, we accept/understand that there is an external object that has some properties and that it can be discretized and counted. But even in the case of the simple one-to-one matching task, we need numbers in order to understand the notion of quantity and to perform correctly this action (Frank et al. 2008: 820; Brannon 2006). The extremely strange Amazonian Pirahã tribe has no words for numbers, or singular–plural morphology. This fact produces severe errors in their judgments about quantities, although it is assumed that they apply approximate magnitude estimations. Words are, then, cognitive tools and language determines Nature and the content of thought (following also the Sapir-Whorf thesis[4]).

(b) Base choice: At the same time, there is a semantic of numerals that must be learned by their users. For example, consider the use of base and the cultural divergences. A look at the history of mathematics shows us the differences, Babylonian used base 60 (sexagesimal), while Mayans base 20, Greeks base 10, Leibniz suggested base 2, a very simple base used later by computers (as well as hexadecimal and, sometimes, octal).

Even in our contemporary decimal system, the sexagesimal (60) base for geometric grades (360) survives like cultural fossils. For time measurements (seconds/minutes) and also the duodecimal (12) system is in use, for example when we buy eggs (e.g., "one dozen, two dozens...."). Even the decimal system is not so strict, when we count 12 months for a year, and the number of days change from month to month. Despite some attempts to change this, like the French Republican Calendar (1793) of 12 months for exactly 30 days with each month one divided into three decades, things remain the same. Concepts are culturally embedded and sometimes survive by tradition, laziness, and disgust toward change. There are more bases: 3-ternary, 4-quaternary, 5-quinary, 6-senary, 7-septenary, 8-octal, 9-nonary, 11-undecimal, 16-hexadecimal ....

---

[4]Quoted by Kay and Kempton (1984), Whorf affirmed in 1956 that "The categories and types that we isolate from the world of phenomena we do not find there because they stare every observer in the face. On the contrary the world is presented in a kaleidoscopic **flux** of impressions which have to be organized in **our** minds. This means, largely, by the linguistic system in our minds." His ideas seemed to emanate two hypotheses: "1st. Structural differences between languages systems will, in general, be paralleled by nonlinguistic cognitive differences, of an unspecified sort, in the native speakers of the two languages; 2nd. The structure of anyone's native language strongly influences or fully determines the world-view he will acquire as he learns the language." By language, we mean also numerical codes.

Consequently, there is no natural way to choose a mathematical base; this reflects a cultural process with some practical points that need our consideration. Perhaps we feel comfortable with base 10 just because most of us have learnt it from our parents, in school and from a variety of social environments that surround us, nothing else. The classic remark that we have 10 fingers and that this is the basis of the decimal system is not strong enough because we know that several cultures make use of hands to calculate relying on several subsections like phalanges or even combining methods to count differently. Counting with hands differs very strongly among European countries and more when compared with Japan, China, or some countries of Africa (Ifrah 1999). The one-to-one counting is not enough to justify the decimal system from the body structure and hand morphology. The uses of the body create numerical meaning. For example, we can explain that in ancient Rome and in the Middle Ages, counting "one" by bending the left little finger, and perhaps the reason for using the left hand was because the right hand was used for more important things (Nishiyama 2010).

(c) Discrete semantics of arithmetic: They take into the consideration the basic operation[5] of $1 + 1 = 2$: while the result is true for two oranges, in the case of the sum of two drops of water this is not true, because the final result is 1 *bigger* drop of water. But even in the first case, we could try to sum oranges and apples and decide that they cannot be summed because of their different Nature. We must learn to create abstract identities that must be classified, accumulated, grouped into classes... and this is not completely obvious or natural. Anybody who has raised a child can observe this very easily. And the empirical studies of Piaget on child mind development, the theory of cognitive development (Piaget 1977), confirms this evidence. Before the preoperational stage, at sensorimotor stage, children have the ability to link numbers to objects, e.g., one dog, two cats, three pigs, four hippos (Ojose 2008), and this trait, found in early development of human cognition, can be modeled correctly with Bayesian ideas (Lee and Sarnecka 2011). Today, Bayesian models produce great models of several cognitive phenomena such as vision, categorization, decision making, language learning, motor control, or a theory of mind (Kwisthout et al. 2011).[6] As a result, there is an ontological stability that

---

[5]This is a not an easy concept. In order to certify the logical necessity and coherence of this arithmetic operation, Bertrand Russell and Norbert Whitehead devoted 379 pages of their *Principia Mathematica* to it (1910–1927); see Volume I, 1st edition, page 379).

[6]These authors provide a good number of bibliographical references about all these domains and the Bayesian modelization. But at the same time, they extend their implementation *ad infinitum* and find a big problem: computational intractability, due to the fact that many of the computations postulated by Bayesian models of cognition fall into the class of NP-hard problems. Would this lead to a bottleneck of intractability once we tried to scale Bayesian basic models to higher cognitive processes? Well, humans are not bottlenecked: We have several decision-making mechanisms. Perhaps at a theoretical level, it is true, in the same way that Gödel dynamited the foundations of mathematics, in front of the horrified Hilbert. But even in that case, they are the best tool we have for several purposes. And it works.

makes it possible to engage in any work with numbers, or at least our success defining some temporary states as stable make it possible.

(d) Finally, there are also psychological constraints: following the study of the human (in) capacity to deal with long lists of numbers, we remember the classic study of Miller (1956) which demonstrated that generally people have problems to remember greater numbers than those of nine units (that is, for example, 798428197). And our necessity to deal with greater numbers has a strong relationship with commerce, war, or even with the time calculation (calendars) and human body regularities. The embodied cognitive inference could explain this, as Zaslavsky (1992) suggests:

> The cyclical Nature of menstruation has played a major role in the development of counting, mathematics, and the measuring of time… Lunar markings found on prehistoric bone fragments show how early women marked their cycles and thus began to mark time. Women were possibly "the first observers of the basic periodicity of Nature, the periodicity upon which all later scientific observations were made" (quoted from Thomson 1981, page 97).

In this later case, for an understanding of a ± 28 day cycle we need a notation system, something beyond intuition and that requires us to take a step forward. The fact is that the analysis of natural cycles, especially astronomical ones, has been the basis of mathematical advances throughout history. We will see in the following chapters that the technical analytic requirements of astronomy led to the birth of modern statistical analysis.

Well, we have been talking about numbers as external entities that must be captured by symbols that have a body correlation. However, another aspect must be considered: the *numbers inside the body*. Beyond any symbolic instantiation of numbers, human bodies perform several calculations through their nervous systems, generating internal probabilities distributions rather than deterministically selected information based on all the available information (Vul and Pashler 2008). In fact, as Körding (2007: 606) points out, "many of the properties of the nervous systems and the bodies of animals are remarkably well adapted to their ecological niche." Neuroscience studies how animals control their limbs and really perform inferences about events in their world and choose among several kinds of rewards. So, the central nervous system (henceforth, CNS) has a basic purpose to make decisions to interact with the environment. Perceptual and motor tasks processed through the CNS are truly probabilistic and must face up to noisy/local/ambiguous sensory cues in a perpetually changing world of new stimulus and facts. And from an evolutionary point of view, the basic necessity of a complex CNS was how to deal properly with movement (Llinás 2001), and establishing connections between motor actions and mentality. For any action, and more so in open and unpredictable environments, there are several possible outcomes that must be considered by the CNS. Daily activities of moving/freezing resting, foraging, fighting/flying involve predictions and computations of action utility functions. Without numbers, the brain performs a continuous statistical activity in order to solve several daily necessities about perception, decision making, and motor control (Doya et al. 2006).[7] From sensory signals (dynamic, multimodal, extended, time varying) to brain neurons and again to body reactions, the whole process of a living entity is a selection process in continuously changing environments. Some Bayesian statistical models have been suggested as possible accurate descriptions of the real mechanism of the spiking neurons (Deneve 2005; Doya et al. 2006) or cognitive processes (Perfors et al. 2011).[8] Nevertheless, some attempt to provide a robust frequentist model for the Nature of this neuronal process has also been published (Martignon and Deco 1997). The truth is that in this research area the debate among Bayesians and frequentists is not so intense as at the philosophical level, because both approaches are useful for different purposes (Bayesian for estimating information transmission while frequentism for detecting spatio-temporal patterns among groups of spiking neurons, for example). The debate is more intense when we discuss the evolutionary mind structure and which model is closer to it: Bayesianism or frequentism (Sloman et al. 2003). Gigerenzer and Hoffrage (1995), first, and echoed later by Cosmides and Tooby (1996)[9] defended a basic "frequentist" approach, affirming that the mind was tuned with frequency formats but at the same time their studies of how information is naturally processed allowed us infer a Bayesian flavor of natural minds. This can be true in certain advanced statistical contexts, because it is obvious that frequentism helps to eradicate some cognitive biases (overconfidence, the conjunction fallacy, and base-rate neglect). This is only a true statement if we take into account the visualization models of data, not the reasoning by itself.[10] Again, the tool determines the usability although this does not imply that the tool is embedded into the mind. Later, Zhu and Gigerenzer (2005) gave support to the notion that most animals are fairly good Bayesians as well as the fact that children use a basic form of "Bayesian reasoning." Again Gigerenzer, in 1991, published an exceptional paper contributing to this very interesting debate "From Tools to Theories: a Heuristic of Discovery in Cognitive Psychology." Here, the cognitive emphasis is made in tools: as soon as a new tool dominates a field, it is considered the "natural" way to perform an action/heuristics, consequently psychological studies take it as a granted model. For example, around 1955 psychophysics studied absolute and differential thresholds in tone recognition. Absolute studies had been the common way to study psychophysics but differential eluded the formal analysis

---

[7]Further: very important notions, such as *surprise*, have intuitive but at the same time quantifiable statistical properties. For a Bayesian account for surprise, see Baldi and Itti (2010).

[8]Some "prior engineering" in robots has been already considered, but the results are still not impressive. Read "the probabilistic mind," by Sanders (2011).

[9]Cosmides and Tooby consider that natural evolved solving-problem algorithms are content dependent and that even Bayesian approaches are content independent. Well, I am not agree directly: Priors allow to be content dependent, although the mechanistic process follows a similar formal pattern. From all the possible ways of dealing with numbers, Bayesian is more adaptive.

[10]Trimmer et al. (2011) suggest the contrary: Bayesian paradigm applied to the evolution of decision-makers allows solving the Ellsberg paradox. The truth is that different models can solve different problems, but this does not credit any of them as better than the others.

because of its then intractability. They created the theory of signal detectability (TSD), using the Neyman-Pearson technique of hypothesis testing, which made it possible to analyze differential psychophysical stimulus. That is, a new tool created a new range of available data, at the time that specialists tended to assume that human minds were frequentists, just because the tool they employed was frequentist. The same happened later when the theories of human causal reasoning (Michotte, Piaget, Gestalt psychologists) that investigated how certain temporospatial relationships between two and more visual objects, such as moving dots, produced phenomenal causality. Then, Harold Kelley proposed his attribution theory in 1967, in the middle of the institutionalization of inferential statistics, and soon the gem of statistical research, Fisher's ANOVA became the fundamental tool of the behavioral scientist. Then, the Fisherian mind appeared. But the truth is that despite the tool we have in mind, human kids use several types of causality approaches (up to 17!). Téglás et al. (2011) have proved that 12-month-old preverbal infants can also make predictions by pure reasoning, that is, probabilistic inferences. Additionally, it has been demonstrated that young children can make approximate comparisons of quantities before knowing the numerical symbolic system (González and Girotto 2011) or infer causes of failed actions, tracking intuitively the statistical dependence between objects, agents, and outcomes (Gweon and Schulz 2011). And their looking times are consistent with a Bayesian ideal observer embodying abstract principles of object motion. Common sense is Bayesian, or the homunculus statistician is Bayesian, rather than frequentist (Gigerenzer 1991: 6).

1.2 Natural Statistics...or Probability Without Numbers

At the First Joint Congress on Evolutionary Biology, held in July 2012 in Canada, evolutionary geneticists Tristan Long, of Wilfrid Laurier University in Waterloo, Canada, and William Rice, of the University of California, Santa Barbara, presented the preliminary results of their ongoing research: after repeatedly subjecting fruit flies to a stimulus (designed to select the presence of numerical skills), a group of evolutionary geneticists finally hit on a generation of flies that could count (it took 40 tries before the species' evolution occurred). They *selected* a race of numerically savvy insects, and this is a successful piece of research on the neuroarchitecture of counting at the same time that shows how numerical skills can emerge by evolution. Nevertheless, this could be considered the result of a direct manipulation of the animals...but what happens in wild Nature? How animals deal with quantities of objects in the middle of dynamic environments under uncertain conditions? How can they decide among several strategies, even in the case of low smart animals? Before showing you some examples from the phylogenetic tree, I would like to introduce two important concepts related to numerosity, defined as "the ability to appreciate and understand numbers" (Coolidge and Overmann 2012): subitization and magnitude appreciation. As we have seen in the example of fruit flies, they have some basic and innate ability to process quantities. Magnitude appreciation is a first step toward the capacity of numerosity. Furthermore, numerosity needs a second skill, *subitization*, the act of immediate recognition of small number of objects, usually within the range of 1–4 tokens. Subitization is not the same as counting, because it is innate and immediate, while the second implies a more elaborate mental process. Numerosity is then an innate skill of several animals and it is shared as a stable trait along the phylogenetic tree, necessary for spatial cognition (even for *blind* living animals, like bats). It becomes the non-symbolic basis for the emergence of symbolic thinking and human infants (9 months old), non-human primates, several mammals, and birds share this numeric ability. Let me to introduce the reader to some direct examples somehow following the phylogenetic tree.

Amoebae. One experiment: slime mold *Physarum polycephalum* exposed to unfavorable conditions presented as three consecutive pulses at constant intervals reduce their locomotive speed in response to each episode. But an even more suggestive result is that when the plasmodia were subsequently subjected to favorable conditions they spontaneously reduced their locomotive speed at the time when the next unfavorable episode would have occurred. Thus, they counted and were able to anticipate their actions (Saigusa 2008).

Ants. Ants need to make measurements. Wittinger et al. (2006) have shown that during ant navigation processes, these insects measure distances traveled using a step integrator or "step counter." Manipulating the lengths of the legs, the scientists were able to appreciate that ants walking on stilts overestimated distances while those walking on stumps underestimated the distances. Here, we find an example of grounded cognition based in morphological traits (Barsalou 2008)[11] as well as with the intuitive (Davies 2004) step counting process.

Fishes. According to Agrillo et al. (2008, 2009), fishes are able to discriminate quantities. It was inferred from their spontaneous tendency to join the largest social group, a very useful skill to survive in open water. To be able to perform magnitude appreciation would then be a cognitive skill directly related to higher survival rates. No subitization has been observed.

Salamanders. From an experiment made by Uller et al. (2003), it was demonstrated that given a choice between tubes containing two fruit flies or three, salamanders lunge at the tube of three. It leads to a notion of number that evolved at least 28 million years ago. Beyond three objects, as we have seen in Sect. 1.1, they fail— as well as monkeys and babies do—and feel themselves confused. This is evident over the length of the susceptible objects to be tracked.

Honeybees. Gross et al. (2009) has shown that honeybees have a sense of numbers, similarly to many mammals. Honeybees training, in a y-maze apparatus and under

---

[11]Barsalou defends statistical processing as central to grounded cognition. He even affirms that Bayesian statistics can be viewed as statistical accounts of the multimodal information stored in the dynamic systems that generate simulations and guide situated action. Memories, explicit as well as implicit, are also cognitively oriented, using natural statistical ways of classification, evaluation, and prediction.

DMTS paradigm using one/two dot patterns, was performed successfully for several days, and they were also able to select a three-pattern over a four-pattern, although they were not able to consistently select a four-pattern over a three-pattern, and higher-number tests were also unsuccessful. The ability of honeybees to discriminate between numbers could easily be of evolutionary benefit. For example, it could serve as directional cues, or aid in foraging behavior.

Newborn chicks. Rugani et al. (2009) reported that newborn chicks appear to add and subtract, consequently proto-arithmetic capacities in the young and relatively inexperienced chicks of this precocial species. This study aimed at extending comparative research on the spontaneous representation of number to very young birds, employing filial imprinting to familiarize the animals with a certain number of elements. These animals did not receive numerical training, but were able to perform basic arithmetic tasks. Here, subitization and magnitude discrimination have added an extra skill: an active way to make "innate calculus."

Dolphins. Recent studies of Yaman et al. (2012) show that dolphins are able to use a numerical category based on "few" versus "many" when discriminating stimuli according to the number of their constituent patterns. At the same time, this study extended the previously demonstrated range of numbers, thereby testing the limits of the numerical abilities of bottlenose dolphins, established now to 6. They also demonstrate that for numbers 1–3 subitization is the basic process followed by dolphins but that from 3 to 6, numerical comparisons are processed logarithmically, as is postulated by the Weber–Fechner law.[12]

The list of animals that are able to perform numerically would be longer, including primates (Thomas et al. 1980; Boysen 1993; Boysen and Hallberg 2000; Smith et al. 2003; Brannon 2006; Jordan and Brannon 2006; van Marle et al. 2006; Addessi et al. 2007; Beran 2007; Cantlon and Brannon 2007; Hanus and Call 2007; Nieder and Merten 2007; Beran et al. 2008), pigeons (Scarf et al. 2012), or new Zealand robins Hunt et al. 2008). The sum of all this evidence makes it possible to affirm that numerosity is something natural previous to abstract thinking that can be traced throughout the evolutionary phylogenetic tree. Now let us take a look at humans.

---

[12]Weber–Fechner law combines two different laws: Weber's law and Fechner law, both interested on quantifying the perception of change in a given stimulus. Weber's law expresses a general relationship between a quantity or intensity of something and how much more needs to be added for us to be able to say that something has been added (this can be studied by discrimination threshold experiments). This general relationship between the initial intensity of something and the smallest detectable increment is exactly what Weber noticed and formalized into "Weber's law." Considering that the relation between one stimulus and its perception is logarithmic, the stimulus changes geometrically and the perception arithmetically. M. S. Landy provides a very elegant mathematical formulation and explanation of both laws in one of his courses' materials, from which the present information has been quoted: http://www.cns.nyu.edu/~msl/courses/0044/handouts/Weber.pdf (Accessed in May 18th 2013).

1.3 The Emergence of Natural Statistics in Humans

When we try to understand the roots of statistical reasoning and its evolutionary necessity, a suggestive question appears in front of us: as we have demonstrated that several kinds of animals perform statistical activities for their own survival, but what can be said about humans? Owing to the difficulties of the absolute blur between culture and Nature, it is still possible to learn something about the innate statistical mechanisms of humans if we choose children as subjects of our analysis.

Even more, we can observe children that have not received language training yet nor can make use of it, and consequently they are not influenced by it.

Young children's number development is a classic domain of research in cognitive sciences (Lee and Sarnecka 2011), and it is a well-established fact, as we have seen in the previous section that humans and other animals have a natural skill to represent small, exact set sizes (up to about 4), and can even make approximate comparisons of quantities before knowing the numerical symbolic system (Barth et al. 2005). They perform approximate arithmetic operations, those necessary to manage their tiny world. Experimental results also support the possibility that linguistic constructions are acquired probabilistically from cognition-general principles (Hsu et al. 2011).

Besides natural counting, there is a second domain in which we must be introduced: causality. Causality is also a necessary knowledge about world functioning,[13] and it is basic for any action: Does this object move? Does this piece hit my foot if I throw it? Is it as good as food? Causality is not only a necessity for high-level abstract knowledge (science is included into this domain) but also for basic day-to-day information necessary for one's own survival. From a basic evolutionary perspective, pain is a good embodied discriminator of negative inputs: for example, pain and the understanding of the mechanisms by which it is produced save us from a young death.[14] 16-Month-Olds rationally infer causes of failed

---

[13]At least for Western thinkers, *cognitio per causas* has been a necessary heuristic in order to understand the world.

[14]Recently, I exposed these ideas as Keynote speaker at EBICC2012 (Brazil), published as "O SIGNIFICADO DO SIGNIFICADO: Novas Abordagens das Emoções e Máquinas", in Gonçalves (2013): UNESP. Familial dysautonomia (FD), also called Riley–Day syndrome, is a genetic inherited disorder that affects the development and function of nerves throughout the body. Among several symptoms, perhaps the most significative for us here is the inability to feel pain and changes in temperature (Rahalkar et al. 2008). This disorder leads easily and fast to death. If we look at the evolution of nociception and the emergence of pain, we can discover very interesting things that show us a new conceptual framework for the analysis of human emotions. First of all is the evolutionary emergence of complexity into nociceptors and nervous systems (Smith and Lewin 2009; Sneddon 2004), which could help us empirically study the elusive Nature of consciousness. Second is the existing similarity from invertebrates right through to humans. Noxious arrays of stimuli (mechanical, thermal, and chemical, from a body perspective but also social, symbolic, or linguistic, in the human case) are threatening forces that any living entity must "understand" to be able to react. Unicellular bacteria such as *E. coli*, although they have no nervous system, they have mechanosensitive channels that make them possible to react to those stimuli. Although they have

actions, as suggested by Gweon and Schulz (2011). From minimal data and sparse evidence, very young children are able to draw accurate inductive inferences and even change their actions when expected outcomes are the result of their "simple" investigations (Cordes and Brannon 2009; Xu 2003). Infants can track the statistical dependence between *objects, agents,* and *outcomes* to perform their initial rational actions (Strauss and Curtis 1981; Xu and Spelke 2000).

Menninger (1992) noted that number words appeared first in writing and only later in speech. And from archaeological evidence, we can infer that number concept was much older than writing and that one-to-one correspondence matching was present within Neolithic farmers in the Middle East (Hyde and Spelke 2008; Piazza et al. 2004). From previous sections, we have shown that subitization is an earlier stage for finger counting, a subjective mode and analogical way to count (creating identity relationships between objects and body sections, Coolidge and Overmann 2012).

As a conclusion, we can affirm that for most animals with spatio-visual tools used to manage their basic interaction with the world, a sense of numerosity is necessary to improve their chances of survival: selecting biggest sources of food, discerning best mating groups, deciding among optimal distances, and so on. The sense of quantities, even in an informal way, makes it absolutely necessary for an interaction with the world. Although there is a space for minimal cognition and even for non-symbolic computing approaches to cognition (like in morphological computing, Casacuberta et al. 2010), even in the case of these minimal systems, a sense of numerosity is embedded and, consequently, the emergence of more sophisticated ways to deal with numbers is something completely coherent, natural, and logical. Even if we look at most recent studies on human cultural evolution (Pagel et al. 2013), the existence of a set of such highly conserved words among seven language families of Eurasia postulated to form a linguistic superfamily that evolved from a common ancestor around 15,000 years ago can be demonstrated.

(Footnote 14 continued)

no true nociceptive response, they have the basis for it. The nervous system, the basic piece of all this emotional arousal, was originated during the early evolution of Eumetazoa (animals with tissues). In some cases, like in Placozoa and Parazoa such as Porifera (sponges), if it is true that they do not have a nervous system, they present genes associated with neural development, something like "proto-neural" cells. It was within bilaterates that the nervous system emerged. It was in the Annelida *Hirudo medicinalis* that nociceptive cells were identified firstly. The sea slug *Tritonia diomedia* shows escape swimming that supports the idea of triggering of nociceptive responses. Nematoda like *Caenorhabditis elegans* and Arthropoda like *Drosophila melanogaster* demonstrate how the evolution of bilateralism enabled a more structured nervous system specialized in the detection of noxious stimuli. Here, we have pain with feeling, because in the previous examples, the nociception activation by itself was not pain. In lower vertebrates, the nociception evolved becoming more specialized, and this was accentuated with Amphibia, Reptilia, Aves, and Mammals. The tree of life offers us an empirical path to the analysis of basic emotions and the possible understanding of the mechanisms that made the emergence of the consciousness possible, always by the hand of emotions (Damasio 1999; Llinás 2001). Consequently, basic emotions are hardwired, even unconsciously as can be shown in reflex acts.

Since the end of the last ice age, some proto-words have survived and among them, we can find…numerals. As a final conclusion, we can affirm that mathematics is panhuman, as ethnomathematicians have demonstrated (Chrisomalis 2009: 495).

1.4   So…Must Statistics Be Considered as a Property of the World or the Result of Our Ignorance?

The Nature of chance has been very controversial all throughout human history, as well of some other ideas as the emptiness (remember the classic *horror vacui*) or the eternity of the world, and it is because there is a strong bond between chance and causality. I will explain it you in a nutshell, and taking as a framework the idea of the (in)existence of chance:

(a) First option: Chance does not exist, and it is only the result of our ignorance about the real causes that rule the world. In a perfect knowledge situation, we could control and understand each one of the variables which are involved into one action and we could easily predict the final outcome. If we cannot do it, and this is the most common situation in our daily life, it is because we do not have a deep knowledge about the real Nature of the objects under our scrutiny neither about the laws that affect to them. Then, everything is the result of a predictable set of variables and consequently we live in a deterministic world, with no choice or freedom. Causality led us to determinism. But, on the other hand,

(b) Second option: if chance does really exist, then there are two consequences:

  (i) First of all, moral/intentional: Chance is the true and core characteristic of our universe, which is not governed by anything nor by anyone. What then is the meaning of our world and life?

  (ii) Second, an ontological problem: Chance exists, but at the same time, deterministic rules exist affecting our universe. Then, it is true that chance exists and, at the same time, it is not the result of our ignorance but a characteristic of the universe. In this case: Is there a reality scalable between determinism and chance? Chance forms the basis of quantum reality or evolutionary biology, while determinism seems to dominate at macro level (objects, planets, galaxies). Heisenberg's uncertainty problem (regarding the uncertain relation between the position and the momentum, or mass times velocity, of a subatomic particle, and the epistemological paradox according to it "the more precisely the position is determined, the less precisely the momentum is known in this instant, and vice versa"[15]) and the Brownian motion (the random drifting of particles suspended in a fluid) are examples of the reality of chance in the

---

[15]English translation from the original text in German (Heisenberg 1927).

government (or lack of any control or determinacy) over the natural events. On the contrary, at higher or macroscopic scales, events are ruled by 4 deterministic elementary forces[16] that alone or combined make it possible to predict conceivable outcomes of events. Therefore, stochastic processes lie at the heart of Nature, although it does not mean that determinism is not possible.

By previous examples, we can affirm that human minds are not naturally ready to deal with mathematical numbers, whatever the low complexity of their range, but at the same time, there is also a natural[17] numerosity skill necessary to work with two kinds of events: first of all, the knowledge related to direct survival (to take care of their breeding, to identify a possible situation of danger by number inferiority and other daily situations); secondly, decision-taking processes in which an intuitive evaluation of previous events is involved, and this implies a rude, basic and natural way to add or subtract positive/negative marked values over previous situations (or in more sophisticated beings, even imagined). As a conclusion, we admit that the use of naïve, intuitive numbers is a reality in the natural world.

Then, several living animals show innate numerosity skills and it is plausible that from this ability emerged a numeric knowledge of the world, as humans do, although I will avoid here the philosophical debate about how the numbers acquired quantitativeness (Wittgenstein 1953). For me it now more important to focus on another aspect: how to solve specific tasks in which a fuzzy idea of number and abstract quantity are involved. This is one of the benefits of future statistical thinking: how to deal appropriately with uncertain or excessively disperse/broad information to give an answer.

Swarm intelligence, like that intelligence showed by ants, is a kind of collaborative task-design in which there are no words or numbers, although a basic chemical language exists. It has been under intense study by AI programmers, trying to obtain benefits from bioinspiration (Bonabeau 1999). Perhaps they decide things, but the amount of information and/or conceptual rules to extract information from it is very limited. Only when a rich environment, and complex tasks related to it must be solved, can statistical thinking be necessary. The truth is looking at Nature we can demonstrate that from irrational individual's group rationality can emerge (Sasaki and Pratt 2011).

---

[16]These four fundamental forces are as follows: strong, electromagnetic, weak, and gravity.

[17]Or prewired, following Page 1, 2012. This work can somehow be situated within the range of the debate on the cognitive Nature of human beings, a topic lead since the last decades by the notion of "Universal Grammar" by linguist Noam Chomsky. For a deeper analysis of this question, see Pinker (1997).

1.5   Then…What the h__ll Is Statistics?

Surely you know the exact letter necessary to understand the previous phrase. You have knowledge enough about language and its most frequent constructions, besides some popular expressions, so it is easy to fill the blank space. Forgive my impoliteness, I was just trying to show you how easily evolutionary meaning of statistics and how it works can be conveyed. We recognize shapes from blurred, fuzzy, evolving, and dark scenarios; our minds are evolutionarily prepared to find meaning of the world. This explains the psychological mechanisms of *pareidolia*, which explains how somebody can find a Jesus face in a slice of toast. This pattern-attentiveness skill was very useful during millennia from a survival perspective. It is an example of several more strategies that emerged during natural evolution to deal with the environment. Those animals with the ability to take good decisions involving several sources of information and possible outcomes had highest ratio of survival and, hence, of disseminating their genes through their offspring.

My starting point is, animals, and from now our animals under study will be humans, are faced with questions for which they do not have enough information. They never can know all the possible data of any object/event (considering also epistemic layer levels), and, consequently, they need to make generalizations from insufficient data. This cognitive situation has been defined as "bounded rationality"[18] and the seminal paper on human real heuristics was written by Tversky and Kahneman (1974). At this point, it was clear that the dual role of statistics as a tool and a model of mind was an established fact. When Edwards, Lindman, and Savage proposed Bayesian statistics as the true way to perform scientific analysis of data, they considered at the same time the mind as a reasonably good, albeit conservative, Bayesian statistician. This is *the hidden legacy that tools bequeath to theories* (Gigerenzer 1991: 11). Besides this fact, and returning to our previous debate, some events, like dice, are intrinsically hazardous. The search of rules, sense, and order into the universe is the response to this necessity. In a first moment, magic answers will appear and, then, scientific truths will illuminate the darkness of mind slowly.

---

[18]Going beyond, and accepting the impossibility of an universal implementation of procedural/logicist rationality in all human action and decision domains, some authors consider bounded rationality as an ecological rationality adapted to the environment features (see Gigerenzera and Todd (1999). They consider that this approach is successful, fast and frugal and escapes from the classic algorithmic approach, defending stronger heuristic thesis. They consider it as an alternative to Bayesian thinking (Albert 2009:63) and defend what they call "cognitive algorithms" (Gigerenzer and Goldstein 1996), as realizations of a framework for modeling inferences from memory, the *theory of probabilistic mental models*.

References

Addessi, E., Crescimbene, L., & Visal-berghi, E. (2007). Do capuchin monkeys (*Cebusapella*) use tokens as symbols? *Proceedings of Biological Sciences, 22*, 2579–2585.

Agrillo, C., Dadda, M., Serena, G., & Bisazza, A. (2008). Do fish count? Spontaneous discrimination of quantity in female mosquitofish. *Animal Cognition, 11*(3), 495–503.

Agrillo, C., Dadda, M. Serena, G., & Bisazza, A. (2009). Use of number by fish. *PLoS ONE, 4*(3): e4786, 1–7.

Albert, M. (2009). Why bayesian rationality is empty, perfect rationality doesn't exist, ecological rationality is too simple, and critical rationality does the job. In M. Baurmann & B. Lahno (Eds.), *Rationality, markets and morals. Perspectives in moral science* (pp. 29–65). Frankfurt, Main: Frankfurt-School-Verlag.

Baldi, P. F., & Itti, L. (2010). Of bits and wows: A bayesian theory of surprise with applications to attention. *Neural Networks, 23*(5), 649–666.

Barsalou, L. W. (2008). Grounded cognition. *Annual Review of Psychology, 59*, 617–645.

Barth, H., La Mont, K., Lipton, L., & Spelke, E. (2005). Abstract number and arithmetic in preschool children. *Proceedings of the National Academy of Sciences, 102*, 14116–14121.

Bender, A., & Beller, S. (2012). Nature and culture of finger counting: Diversity and representational effects of an embodied cognitive tool. *Cognition, 124*, 156–182.

Beran, M. J. (2007). Rhesus monkeys (*Macacamulatta*) enumerate large and small sequentially presented sets of items using analog numerical representations. *Journal of Experimental Psychology: Animal Behavior Processes, 33*, 42–54.

Beran, M. J., Evans, T. A., Leighty, K. A., Harris, E. H., & Rice, D. (2008). Summation and quantity judgments of sequentially presented sets by capuchin monkeys (*Cebusapella*). *American Journal of Primatology, 70*, 191–194.

Bonabeau, E., Theraulaz, G., & Dorigo, M. (1999). *Swarm intelligence: From natural to artificial systems (Santa Fe Institute Studies in the Sciences of Complexity)*. UK: OUP.

Boysen, S. T. (1993). Counting in chimpanzees: Non human principles and emergent properties of number. In S. T. Boysen & E. J. Capaldi (Eds.), The development of numerical competence: Animal and human models (pp. 39–59). Hillsdale, NJ: Lawrence Erlbaum Associates.

Boysen, S. T., & Hallberg, K. I. (2000). Primate numerical competence: Contributions toward understanding non human cognition. *Cognitive Sciences, 24*, 423–443.

Brannon, E. M. (2006). The representation of numerical magnitude. *Current Opinion in Neurobiology, 16*, 222–229.

Cantlon, J. F., & Brannon, E. M. (2007). How much does number matter to a monkey (*Macaca mulatta*)? *Journal of Experimental Psychology: Animal Behavior Processes, 33*, 32–41.

Carsetti, A. (Ed.). (2010). *Causality, meaningul complexity and embodied cognition*. Heidelberg: Springer.

Casacuberta, D., Ayala, S., & Vallverdú, J. (2010). Embodying cognition: A morphological perspective. In J. Vallverdú (Ed.), Thinking machines and the philosophy of computer science: Concepts and principles (pp. 344–366). USA: IGI Global Group (Editor i autor).

Chrisomalis, S. (2009). The cognitive and cultural foundations of numbers. In E. Robson & J. Stedall (Eds.), *The oxford handbook of the history of mathematics* (pp. 495–517). Oxford: Oxford University Press.

Coolidge, F. L., & Overmann, K. A. (2012). Numerosity, abstraction, and the emergence of symbolic thinking. *Current Anthropology, 53*(2), 204–225.

Cordes, S., & Brannon, E. M. (2009). Crossing the divide: Infants discriminate small from large numerosities. *Developmental Psychology, 45*, 1583–1594.

Cosmides, L., & Tooby, J. (1996). Are humans good intuitive statisticians after all? Rethinking some conclusions from the literature on judgment under uncertainty. *Cognition, 58*(1), 1–73.

Damasio, A. (1999). *The feeling of what happens*. London: Heinemann.

Davies, W. M. (2004). Amodal or perceptual symbol systems: A false dichotomy? *Behavioral and Brain Sciences, 27*(01), 162–163.

Deneve, S. (2005). Bayesian inference in spiking neurons. In L. K. Saul, Y. Weiss, & L. Bottou (Eds.), *Advances in neural information processing systems 17* (pp. 353–360). USA: MIT Press.

Derbyshire, D. C., & Pullum, G. K. (Eds.). (1986). *Handbook of amazonian languages* (1st ed.). Berlin: Mouton de Gruyter.

Doya, K., Ihsii, S., Pouget, A., & Rao, R. P. N. (Eds.). (2006). *Bayesian brain: Probabilistic approaches to neural coding.* Cambridge: The MIT Press.

Frank, M. C., Everett, D. L., Fedorenko, E., & Gibson, E. (2008). Number as a cognitive technology: Evidence from Pirahã language and cognition. *Cognition, 108*(3), 819–824.

Gigerenzer, G. (1991). From tools to theories: A heuristic of discovery in cognitive psychology. *Psychological Review, 98*(2), 257–267.

Gigerenzer, G., & Goldstein, D. G. (1996). Reasoning the fast and frugal way: Models of bounded rationality. *Psychologicl Review, 103*(4), 650–669.

Gigerenzer, G., & Hoffrage, U. (1995). How to improve bayesian reasoning without instructions: Frequency formats. *Psychological Review, 102*(2), 684–704.

Gigerenzer, G., & Todd, P. M. (1999). Ecological rationality: the normative study of heuristics. In G. Gigerenzer & P. M. Todd, The ABC Research Group (eds.), *Ecological rationality: Intelligence in the world* (pp. 487–497). New York: Oxford University Press.

Gonçalves, J. (2013). *Encontro com as Ciencias Cognitivas* (Vol. 6). Brazil: Cognição, Emoção e Ação.

González, M., & Girotto, V. (2011). Combinatorics and probability: Six- to ten-year-olds reliably predict whether a relation will occur. *Cognition, 120*, 372–379.

Gordon, P. (2004). Numerical cognition without words: Evidence from Amazonia. *Science, 306* (5695), 496–499.

Gross, H. J., Pahl, M., Si, A., Zhu, H., Tautz, J., & Zhang, S. (2009) Number-based visual generalization in the honeybee. *PLoS ONE, 4* (1), e4263, 1–9.

Gvozdanovic, J. (2006). Numerals. *Keith Allan, concise encyclopedia of semantics* (pp. 736–739). Amsterdam: Elsevier Science.

Gweon, H., & Schulz, L. (2011). 16-Month-Olds rationally infer causes of failed actions. *Science, 332*, 1524.

Hanus, D., & Call, J. (2007). Discrete quantity judgments in the great apes (Panpaniscus, Pantroglodytes, Gorillagorilla, Pongopygmaeus): The effect of presenting whole sets versus item-by-item. *Journal of Comparative Psychology, 121*, 241–249.

Heisenberg, W. (1927). Über den anschaulichen Inhalt der quantentheoretischen Kinematik und Mechanik. *Zeitschrift für Physik, 43*(3–4), 172–198.

Hsu, A. S., Chater, N., & Vitányi, P. M. B. (2011). The probabilistic analysis of language acquisition: theoretical computational and experimental analysis. *Cognition, 120*, 380–390.

Hunt, S., Low, J., & Burns, K. C. (2008). Adaptive numerical competency in a food-hoarding songbird. *Proceedings of Biological Sciences, 275*, 2373–2379.

Hyde, D. C., & Spelke, E. S. (2008). All numbers are not equal: An electrophysiological investigation of small and large number representations. *Journal of Cognitive Neuroscience, 21*, 1039–1053.

Ifrah, G. (1999). *The universal history of numbers: From prehistory to the invention of the computer.* UK: Wiley.

Jordan, K. E., & Brannon, E. M. (2006). A common representational system governed by Weber's law: Non verbal numerical similarity judgments in 6-year-olds and rhesus macaques. *J. Exp. ChildPsychol., 95*, 215–229.

Kay, P., & Kempton, W. (1984). What is the Sapir-Whorf Hypothesis? *American Anthropologist, 86*(1), 65–79.

Körding, K. (2007). Decision theory: What "Should" the nervous system do? *Science, 318*, 606–610.

Kwisthout, J., Wareham, T., & van Rooij, I. (2011). Bayesian intractability is not an ailment that approximation can cure. *Cognitive Science, 35*, 779–784.

Lee, M. D., & Sarnecka, B. W. (2011). Number-knower levels in young children: Insights from Bayesian modeling. *Cognition, 120*, 391–402.

Llinás, R. R. (2001). *I of the vortex, from neurons to self.* Cambridge, MA: MIT Press.

Martignon, L., & Deco, G. (1997). *Detecting spatio-temporal patterns among groups of spiking neurons: A frequentist approach.* Breckenridge, USA: Workshop on Neurostatistics and Cell Assemblies.

Menninger, K. (1992). *Number words and number symbols: A cultural history of numbers.* New York: Dover.

Miller, G. A. (1956). The magical number seven, plus or minus two: Some limits on our capacity for processing information. *The Psychological Review, 63,* 81–97.

Nieder, A., & Merten, K. (2007). A labeled-line code for small and large numerosities in the monkey prefrontal cortex. *Journal of Neuroscience, 27,* 5986–5993.

Nishiyama, Y. (2010). Counting with the fingers. http://www.osaka-ue.ac.jp/zemi/nishiyama/math2010/finger.pdf. Accessed in August 12, 2014.

Ojose, B. (2008). Applying Piaget's theory of cognitive development to mathematics instruction. *The Mathematics Educator, 18*(1), 26–30.

Pagel, M., Atkinson, Q. D., Calude, A. S., & Meade, A. (2013). Ultraconserved words point to deep language ancestry across Eurasia. *PNAS* 2013, Published ahead of print May 6, 2013, doi: 10.1073/pnas.1218726110

Perfors, A., Tenenbaum, J. B., Griffiths, T. L., & Xu, F. (2011). A tutorial introduction to bayesian models of cognitive development. *Cognition, 120*(3), 302–321.

Piaget, J. (1977). *Epistemology and psychology of functions.* Dordrecht, Netherlands: D. Reidel Publishing Company.

Piazza, M., Izard, V., Pinel, P., Le Bihan, D., & Dehaene, S. (2004). Tuning curves for approximate numerosity in the human intraparietal sulcus. *Neuron, 44,* 547–555.

Pinker, S. (1997). *The language instinct.* USA: William Morrow & Company.

Rahalkar, MD., Rahalkar, AM., Joshi, SK., (2008). Case series: Congenital insensitivity to pain and anhidrosis. *The Indian journal of radiology & imaging, 18,* 132–134.

Rugani, R., et al. (2009). Arithmetic in newborn chicks. *Proceedings of the Royal Society B, 276,* 2451–2460.

Saigusa, T., et al. (2008). Amoebae anticipate periodic events. *Physical Review Events, 100* (1):018101–1/4.

Sanders, L. (2011). The probabilistic mind. *Science News, 180*(8), 18.

Sasaki, T., & Pratt, S. (2011). Emergence of group rationality from irrational invidivuals. *Behavioural Ecology, 22,* 276–281.

Scarf, D., Hayne, H., & Colombo, M. (2012). Pigeons on par with primates in numerical competence. *Science, 23,* 1664.

Sloman, S. A., et al. (2003). Frequency illusions and other fallacies. *Organizatorial Behavior and Human Decision processes, 91,* 296–309.

Smith, E. S. J., & Lewin, G. R. (2009). Nociceptors: A phylogenetic view. *Journal of Comparative Physiology A, 195,* 1089–1106.

Smith, B. R., Piel, A. K., & Candland, D. K. (2003). Numerosity of a socially housed hamadryas baboon (*Papiohamadryas*) and a socially housed squirrel monkey (*Saimiri sciureus*). *Journal of Computer Psychology, 117,* 217–225.

Sneddon, L. U. (2004). Evolution of nociception in vertebrates: Comparative analysis of lower vertebrates. *Brain Research Reviews, 46,* 123–130.

Strauss, M. S., & Curtis, L. E. (1981). Infant perception of numerosity. *Child Development, 52,* 1146–1152.

Téglás, E., et al. (2011). Pure reasoning in 12-Month-Old infants as probabilistic inference. *Science, 332,* 1054–1059.

Thomas, R. K., Fowlkes, D., & Vickery, J. D. (1980). Conceptual numerousness judgments by squirrel monkeys. *American Journal of Psychology, 93,* 247–257.

Thompson, W. I. (1981). *The time falling bodies take to light.* UK: St. Martin's Press.

Trimmer, P. C., et al. (2011). Decision-making under uncertainty: Biases and Bayesians. *Animal Cognition, 14,* 465–476.

Tversky, A., & Kahneman, D. (1974). Judgement under uncertainty: Heuristics and biases. *Science, 185*(4157), 1124–1131.

Uller, C., Jaeger, R., Guidry, G., & Martin, C. (2003). Salamanders (*Plethodon cinereus*) go for more: Rudiments of number in an amphibian. *Animal Cognition, 6,* 105–112.

Vallverdú, J. (2010). Seeing for knowing. The Thomas effect and computational science. In J. Vallverdú, (ed.), *Thinking machines and the philosophy of computer science: Concepts and principles* (pp. 140–160). Hershey: IGI Global Group.

Van Marle, K., Aw, J., McCrink, K., & Santos, L. (2006). How capuchin monkeys (*Cebusapella*) quantify objects and substances. *Journal of Comparative Psychology, 120,* 416–426.

Vul, E., & Pashler, H. (2008). Measuring the crowd within. Probabilistic representations within individuals. *Psychological Science, 19*(79), 645–647.

Wittgenstein, L. (1953). *Philosophical investigations.* UK: Blackwell.

Wittinger, M., Wehner, R., & Wolf, H. (2006). The ant odometer: Stepping on stilts an stumps. *Science, 312,* 1965–1967.

Xu, F. (2003). Numerosity discrimination in infants: Evidence for two systems of representations. *Cognition, 89,* B15–B25.

Xu, F., & Spelke, E. S. (2000). Large number discrimination in 6-month old infants. *Cognition, 74,* b1–b11.

Yaman, S., Kilian, A., von Fersen, L., & Güntürkün, O. (2012). Evidence for a numerosity category that is based on abstract qualities of "Few" vs. "Many" in the bottlenose dolphin (*Tursiops truncatus*). *Frontiers in Psychology*, November 7, 2012, http://t.co/P0btCsCw

Zaslavsky, C. (1992). Women as first mathematician. International Study Group on Ethnomathematics Newsletter, 7(1), 1.

Zhu, L., & Gigerenzer, G. (2005). Children can solve bayesian problems: The role of representation in mental computation. *Cognition, 98,* 287–308.

Chapter 2
Ancient Statistics History in a Nutshell

Abstract This chapter is entirely devoted to the historical roots of human interest of chance, from gambling activities and theories to the role of insurance companies, the medical aspects of vaccination, the notion of legal evidence or even the existence and intervention of some divinity. Conjoined to several human practices, the philosophical aspects implied into the notion of hazard are explored which were found initially in gambling or predictive rituals, most of them from a religious context.

Keywords Dice · Insurance companies · Lottery · Determinism · Gambling · Astronomy · Luck · Destiny · Wager · Vaccines · Legal · Evidence

At the beginning of the eleventh century, Japanese Emperor Shirakawa[1] recited his own list of three unmanageable things: sōhei (armed monks), dice, and the water of the Kamo River. Those monks died centuries ago and Kamogawa was finally domesticated by modern disciplined engineers, but dice… are still a proof of daily surprise and the simplest example of the reality of chance. They are small toys for fun, sometimes portrayers of good or bad luck, but always a solid geometrical form with some signs painted or grabbed. Dice have fascinated us for several centuries, always changing sides without a reasonable way of knowing their next behavior. They will show us the pathway to statistics, be patient.

2.1 Dice in a Deterministic World

Five thousand years ago, dice were invented in India (David 1998). This fact implies that their users had at least a common sense approach to the idea of probability. Those dice were not the contemporary cubical standard dice, but fruit

---

Some of the data in this chapter has been extracted from my previous research (Vallverdú 2011a, b).

[1]白河天皇, Shirakawa-tennō, July 7, 1053—July 24, 1129, was the 72nd emperor of Japan.

J. Vallverdú, *Bayesians Versus Frequentists*,
SpringerBriefs in Statistics, DOI 10.1007/978-3-662-48638-2_2

stones or animal bones (Dandoy 2006). They must surely have been used for fun and gambling as well as for fortunetelling practices. The worries about the future and the absurd idea that the world was causally guided by supernatural forces led those people to a belief in the explanatory power of rolling dice. In fact, cosmogonical answers were the first attempt to explain in a causal way the existence of things and beings. The Greek creation myth involved a game of dice between Zeus, Poseidon, and Hades. And in the classic Hindu book *Mahabharata* (section "Sabha-parva"), we can find the use of dice for gambling, where it is explained how the Pandavas were robbed of their kingdom by means of a game of dice, but in both cases, there is no theory regarding probability[2] in dice, just their use "for fun." At this point, it is nonetheless necessary to make a stop and take a tour into Indian culture.

According to Raju (Gabbay et al. 2011: 1174–1195), the permutational and combinational theories necessary to calculating probabilities in games of chance, such as dice or cards, were born in ancient India. First in Vedic metre conception and secondly in the Jain *Vyākhyāprajñapti*, commonly known as *Bhagavati sūtra*, the fifth of the 12 sacred āagam or janinist canonical texts. These texts were written by Mahavira's disciples who memorized and transmitted his ideas orally until they were fixed by writing them into books (specifically, *Sutras*) around 4th or 3rd BCE. There, permutations are called *vikalpa-ganita* (the calculus or alternatives) and combinations *bhanga*. While Greek and Roman mathematics were still slaves of a bad notational system, Indians had a perfect place value system and the number zero, something that made it possible to work with very big numbers. It is true that even before, in Egyptian mathematics of 12th Dynasty (ca 1990–1800 BCE) existed even a number for one million drawn as a god with his hands raised in adoration pictogram,[3] and we know that Egyptians worked with high numbers as a consequence of their big governmental necessities (food, prisoners, soldiers,...). Nonetheless even in that case their numbers' size was nothing compared to Indian ones: Jain literature typically runs into very large numbers as 10^{12}, 10^{53}, or even 10^{60}. Large numbers, beyond their magical meaning, demand a use of probability. On the other hand, dice games in India are popular and have attracted intellectual interest. For example, in the *Rigveda*, *Mandala* (book) 10, chapter 34,[4] we find several references to the gambling practices. Then, Indians have at the same time knowledge about dice (and that they are always loaded), and fair/deceitful gambling. This later introduces us to a notion of knowledge of large numbers where the notion of convergence can be found and, even, the foundations or probability theory.[5] We note this and leave the debate at this point, more interesting for

---

[2]For a very detailed history of probability and how the empire of chance emerged among several disciplines, see Gigerenzer, Gerd et al. (1989).

[3]Information obtained from Burton (2005).

[4]Although not written until fourth or sixth century of our era, *Rigveda* was much more ancient.

[5]As attempted to justify in the 1950s P.C. Mahalanobis, J.B.S. Haldane and D.S. Kothari. Raju (2011): 1191 from the 3-valued logic present in Jainism.

specialists in history of mathematics than us, who are following the path from natural numbers to human ways to design strategies to deal with them. Anyhow, numbers are not free from conceptual frameworks from which they emerge.

Again in Western territories, we can consider Aristotle as the strongest defender of the causal and empirical approach to reality (*Physics*, II, 4–6) although he considered the possibility of chance, especially the problem of the game of dice (*On Heavens*, II, 292a30) and probabilities implied in it. These ideas had nothing to do with those about atomistic chance by Leucippus, Democritus[6], or Lucrecius' controversial *clinamen*'s theory. Hald (1988, Sect. 3.2.) affirms the existence of mathematical rather than statistical thought in Classical Antiquity, surely due to the imperfection of the used randomizers (bones of hooved animals instead of regular dice), something that made an axiomatization of games of chance impossible; regardless, we can accept that some authors (like Aristotle) were worried about the idea of chance (as well as about the primordial emptiness and other types of conceptual *cul-de-sac*), but they made no formal analysis of it. Anyhow, trust in order and regularities was the aim of life and philosophy, as we can find in verse 490 of Book 2 of the *Georgics* (29 BC), by the Latin poet Virgil: "Felix qui potuit rerum cognoscere causam" (translated as "Can he happy who is able to know the causes of things").

Later, we can find traces of interest in the moral aspects of gambling with dice in Talmudic (*Babylonian Talmud, Book* 8: *Tract Sanhedrin, chap.* 3, *Mishnas* I *to* III) and Rabbinical texts, and we know that in 960, Bishop Wibolf of Cambrai calculated 56 diverse ways of playing with three dice. *De Vetula*, a Latin poem from the thirteenth century, tells us of 216 possibilities. But the first occurrence of combinatorics per se arose from Chinese interest in future prediction through the 64 hexagrams of the *I Ching* (previously eight trigrams derived from four binary combinations of two elemental forces, *yin* and *yang*). The idea of making combinations in order to obtain several results and find the best options was also described by the Catalan philosopher Raimon Llull in his *Ars Magna* [Ars Maior (1273–74), *Ars inventiva* (1289), and *Ars generalis* (1308)], later updated and improved by Wilhelm Leibniz in his *Dissertatio de arte combinatorial* (1666). What Llull tried to design was a method to convince Muslims about their fundamental error and to demonstrate the "evidence" of the Christian truth. With his conceptual wheels, Llull embraced as a real polymathist all the wisdom of his era: "arbor scientiae," "arbor elementalis," "arbor vegetalis," "arbor moralis," "arbor aspostocalis," "arbor coelestialis," "arbor christianalis," "arbor divinalis," "arbor naturalis et logicalis." It can look like the classic encyclopedism of the Middle Age, but Llull tried in fact to surpass this with a new heuristic of knowledge generation.

Despite the religious flavor of Llull's attempts, it was a Muslim who started the history of statistics, and beyond any religious framework. Instead of being worried

---

[6]As an exception of a whole Western paradigm, however, we find this point in one of the conserved fragments of Democritus: "Everything existing in the universe is the fruit of chance and necessity." A whole recompilation can be found in the classic Diels (1903). Demokritos. A very good paper on this topic is Edmunds (1972).

by proselytism, Abu Yūsuf Ya'qūb ibn' Isḥāq aṣ-Ṣabbāḥ al-Kindī, usually known by Western historians as Al-Kindi (801–873), gave a detailed description of how to use statistics and frequency analysis to decipher encrypted messages. With his book *Manuscript on Deciphering Cryptographic Messages*, he gave birth to both statistics and cryptanalysis. The truth is that Al-Kindi, while working at al-Ma'mun's House of Wisdom (together with al-Khwarizmi and the Banu Musa brothers!) was faced with religious confrontations among orthodox factions, although he was always a neutral philosopher and scientist more interested in general knowledge than religious discussions. Anyway, cryptography had connected politics and numbers or letters and now, statistical approaches had started to change the way by which secrets could be transmitted.

In 1494, Luca Paccioli defined the basic principles of algebra and multiplication tables up to 60×60 in his book *Summa de arithmetica, geometria, proportioni e proportionalita*. He posed the first serious statistical problem of two men playing a game called "balla," which is to end when one of them has won six rounds. However, when they stop playing A has only won five rounds and B three. How should they divide the wager? It would be another 200 years before this problem was solved. In 1545, Girolamo Cardano wrote the books *Ars magna* (the great art) and *Liber de ludo aleae* (the book on games of chance). This was the first attempt to use mathematics to describe statistics and probability and accurately describe the probabilities of throwing various numbers with dice. Galileo expanded on this by calculating probabilities using two dice, writing a small text in 1620, *Sopra le scoperte dei dadi* (*Concerning an Investigation on Dice*). At the same time, the measurement and quantification of all aspects of daily life (art, music, time, space) between the years 1250 and 1600 made possible the numerical analysis of nature and, consequently, the discovery of the distribution of events and their rules (Crosby 1996). It was finally Blaise Pascal who refined the theories of statistics and, later, Pierre de Fermat solved the "balla" problem of Paccioli (Devlin 2008). All these paved the way for modern statistics, which essentially began with the use of actuarial tables to determine insurance for merchant ships (Hacking 1984, 1990). Pascal was also the first to apply probability studies to the theory of decision, curiously, in the field of religious decisions. Despite the previous affirmation, and according to Bellhouse (1988: 63) the beginning of probability began in 1645, it was the time of the Pascal–Fermat correspondence in the middle of puritan casuistry. Puritans were faced with the problem of conciliate to the presence of divination and gambling in the Bible with the notion of God's will into a deterministic world. There is a long list of examples of such events in the Bible,[7] and crucial

---

[7]Bellhouse (1988): 66 quotes Acts 1:23–26; Luke 1:9–11; Matthew 27:35–37; Mark 15:22–24; Luke 23:35, John 19:23–24...but there is a long list of false prophets (Deut. 18:10, 14; Micah 3:6, 7, 11), of necromancers (1 Sam. 28:8), of the Philistine priests and diviners (1 Sam. 6:2), of Balaam (Josh. 13:22). Three kinds of divination are mentioned in Ezek. 21:21, by arrows, consulting with images (the teraphim), and by examining the entrails of animals sacrificed. The practice of this art seems to have been encouraged in ancient Egypt. Diviners also abounded among the aborigines of Canaan and the Philistines (Isa. 2:6; 1 Sam. 28). At a later period,

Christian theologists like Thomas Aquinas devoted part of their researches to battle against chance falls. In the case of Aquinas, he expressed in his *Summa Theologiae* that chance events were part of contingent events, and thus they were far from the true and definitive nature of divine necessary events. Providence is thus deterministic. Finally, in 1662, John Graunt published his mortality tables that produced what has been called "empirical statistics."

It is in this historical moment that the Latin term "probabilis" acquires its actual meaning evolving from "worthy of approbation" to "numerical assessment of likelihood on a determined scale" (Moussy 2005). In fact, Pascal introduced a new concept: the moral wager.

2.2 From Dice to Moral Wagers and God in Mathematics

In 1669, seven years after his death, Blaise Pascal's book *Pensées* was published posthumously. At the beginning of the Third Section, aforism §233, it reads:

"(…)Let us then examine this point, and say, "God is, or He is not." But to which side shall we incline? Reason can decide nothing here. There is an infinite chaos which separated us. A game is being played at the extremity of this infinite distance where heads or tails will turn up. What will you wager? According to reason, you can do neither the one thing nor the other; according to reason, you can defend neither of the propositions.

Do not then reprove for error those who have made a choice; for you know nothing about it. "No, but I blame them for having made, not this choice, but a choice; for again both he who chooses heads and he who chooses tails are equally at fault, they are both in the wrong. The true course is not to wager at all."

(Footnote 7 continued)

multitudes of magicians poured from Chaldea and Arabia into the land of Israel and pursued their occupations (Isa. 8:19; 2 Kings 21:6; 2 Chr. 33:6). This superstition widely spread, and in the time of the apostles there were "vagabond Jews, exorcists" (Acts 19:13), and men like Simon Magus (Acts 8:9), Bar-jesus (13:6, 8), and other jugglers and impostors (19:19; 2 Tim. 3:13). Every species and degree of this superstition was strictly forbidden by the Law of Moses (Ex. 22:18; Lev. 19:26, 31; 20:27; Deut. 18:10, 11). But beyond these various forms of superstition, there are instances of divination on record in the Scriptures by which God was pleased to make known his will. (1) There was divination by lot, by which, when resorted to in matters of moment, and with solemnity, God intimated his will (Josh. 7:13). The land of Canaan was divided by Lot (Num. 26:55, 56); Achan's guilt was detected (Josh. 7:16–19), Saul was elected as king (1 Sam. 10:20, 21), and Matthias chosen to the apostleship, by the solemn Lot (Acts 1:26). It was thus also that the scape-goat was determined (Lev. 16:8–10). (2) There was divination by dreams (Gen. 20:6; Deut. 13:1, 3; Judg. 7:13, 15; Matt. 1:20; 2:12, 13, 19, 22). This is illustrated in the history of Joseph (Gen. 41:25–32) and of Daniel (2:27; 4:19–28). (3) By divine appointment, there was also divination by the Urim and Thummim (Num. 27:21), and by the ephod. (4) God was pleased sometimes to vouch-safe direct vocal communications to men (Deut. 34:10; Ex. 3:4; 4:3; Deut. 4:14, 15; 1 Kings 19:12). He also communed with men from above the mercy-seat (Ex. 25:22), and at the door of the tabernacle (Ex, 29:42, 43). (5) Through his prophets, God revealed himself and gave intimations of his will (2 Kings 13:17; Jer. 51:63, 64). From: Divination. (n.d.). Easton's 1897 *Bible Dictionary*.

Yes; but you must wager. It is not optional. You are embarked. Which will you choose then? Let us see. Since you must choose, let us see which interests you least: You have two things to lose, the true and the good; and two things to stake your reason and your will, your knowledge and your happiness; and your nature has two things to shun, error and misery. Your reason is no more shocked in choosing one rather than the other, since you must of necessity choose. This is one point settled. But your happiness? Let us weigh the gain and the loss in wagering that God is. Let us estimate these two chances. If you gain, you gain all; if you lose, you lose nothing. Wager, then, without hesitation that He is.—"That is very fine. Yes, I must wager; but I may perhaps wager too much."—Let us see. Since there is an equal risk of gain and of loss, if you had only to gain two lives, instead of one, you might still wager."[8]

This colloquial style scandalized his contemporaries as well as posterior thinkers: faith could be rational or not (the classic debate of Middle age), but never be the result of a wager, because this act joined ignominiously the fields of religion and games. This idea will be destroyed by the powerful Kantian moral Metaphysics (*Grundlegung zur Metaphysik der Sitten*, 1785) and not will change until the radical works of Friedrich Nietzsche at the end of nineteenth century and the antifundamentalist ethical advances in the twentieth century. Although Kant also made an indirect reference to a bet as a way to understand whether the things in which we trust are solid enough,[9] he was far from the belief in the presence of chance into

---

[8]Quoted from http://www.gutenberg.org/files/18269/18269-h/18269-h.htm#SECTION_III, accessed May 28, 2013.

[9]"For the subjective grounds of a judgement, such as those that produce belief, cannot be admitted in speculative inquiries, inasmuch as they cannot stand without empirical support and are incapable of being communicated to others in equal measure. But it is only from the practical point of view that a theoretically insufficient judgement can be termed belief. Now the practical reference is either to skill or to morality; to the former, when the end proposed is arbitrary and accidental, to the latter, when it is absolutely necessary. If we propose to ourselves any end whatever, the conditions of its attainment are hypothetically necessary. The necessity is subjectively, but still only comparatively, sufficient, if I am acquainted with no other conditions under which the end can be attained. On the other hand, it is sufficient, absolutely and for every one, if I know for certain that no one can be acquainted with any other conditions under which the attainment of the proposed end would be possible. In the former case my supposition—my judgement with regard to certain conditions—is a merely accidental belief; in the latter it is a necessary belief. The physician must pursue some course in the case of a patient who is in danger, but is ignorant of the nature of the disease. He observes the symptoms, and concludes, according to the best of his judgement, that it is a case of phthisis. His belief is, even in his own judgement, only contingent: another man might, perhaps come nearer the truth. Such a belief, contingent indeed, but still forming the ground of the actual use of means for the attainment of certain ends, I term Pragmatical belief. The usual test, whether that which any one maintains is merely his persuasion, or his subjective conviction at least, that is, his firm belief, is a bet. It frequently happens that a man delivers his opinions with so much boldness and assurance, that he appears to be under no apprehension as to the possibility of his being in error. The offer of a bet startles him, and makes him pause. Sometimes it turns out that his persuasion may be valued at a ducat, but not at ten. For he does not hesitate, perhaps, to venture a ducat, but if it is proposed to stake ten, he immediately becomes aware of the possibility of his being mistaken—a possibility which has hitherto escaped his observation. If we imagine to ourselves that we have to stake the happiness of our whole life on the truth of any proposition, our judgement drops its air of triumph, we take the alarm, and discover the actual strength of our belief. Thus pragmatical belief has degrees, varying in proportion to the interests at stake."

moral sphere. Even in the natural domain, scientists like Einstein declared in the twentieth century that "God does not play dice." There is a cultural horror toward the idea of the chance in the universe, as a rule of the destiny. Nonetheless, the birth of moral statistics and the relationship between God and numbers must still to be explained. Let's go!

In the year of 1612, a big first prize was obtained by the winner of the London Lottery. This Lottery was organized by King James I to obtain funds to help to the colonies established in Virginia.[10] At the same time, he granted to the Virginia Company of London the right to raise money to help establish those settlers in the first permanent English colony at Jamestown (Virginia). Lotteries were created in China in the third to second century BC, in order to obtain funds to run politic activities (financial aid to State projects, pay armies…). 46 years had passed after the first lottery in England, authorized by Queen Elizabeth I, when there was a debate on some social problems on chance games, so it is normal that intellectuals devoted themselves to its study. The French Calvinist Lambert Daneau was the first (in 1566, *Deux traittez de S.c. Cyprian. L'un, contre les ieux ei iouers de cartes & de dez. Le tout mis en francois par L. Daneau*) to write about gambling and religion and suggested which games should be allowed or forbidden to Christians: to the first section belong games of pure chance, while in the second corresponded games of mixed chance and skill. The first English Puritan to write about this topic was Northbrooke (1577).

Some decades later, in 1619, Thomas Gataker published *Of the Nature and Use of Lots*, offering a historical review on games in which hazard is involved, as well as a religious interpretation of Chance. From his own words, in Chap. 2, §1, 6, 7[11]:

> Now because Chance or Casualty bears much sway in Lottery, Casual Events being the subject matter of Lots, the due consideration thereof will help not a little to the clearing of the nature of Lots and Lottery, and those Questions that are moved concerning the same. Concerning Chance therefore or Casualty we will consider four things: (1) the name of it; (2) the nature of the thing so named; (3) two distinct Acts concurring in it, and (4) and lastly, certain conclusions or aphorisms concerning it. (…)By the means whereof it comes oft to pass, the same events are casual to some that foresaw them not, and yet not casual to others that foresaw them before. And so it is true, that Casualty depended upon our ignorance; which therefore the more we know, the less we are subject unto. §7. And hence follows the fourth and last Conclusion: *That there is no casualty with God, because there is no ignorance in God.* There is nothing, I say, casual unto Him; nothing comes contingently, but all things are necessarily in regard of Him and His decree."

(Footnote 9 continued)

Critique of Pure Reason, A825/B853. Quoted from http://www.gutenberg.org/files/4280/4280-h/4280-h.htm, accessed in May 28th 2013. In 20th Century Bruno de Finetti will offer a more sophisticated version of Kant's approach to the confidence evaluation of own opinions.

[10]Only in 1612, the benefits of this lottery amounted to nearly of £30.000, according to Holmes (1826).

[11]The full edited text can be found at http://www.conallboyle.com/lottery/GatakerNature_UseofLots.pdf, accessed in May 29, 2013.

Here, and italics in the previous text are mine, we find one of the most current ideas of ancient thinkers: casualty (or chance or hazard) is nothing but the result of human ignorance; God knows everything and, then, for him (yes it is a *he*), there is no casualty. Chance is a consequence of ignorance, not a real dimension of the reality, just a mistake emerged from human fuzzy cognition.[12] Consequently, childish bibliomantic practices were banned by early Christians, although not very successfully: among the Christians remained some fortune-telling practices, as Bible lottery or *sortes Biblicae*, a method consisting by taking random passages from the Bible and to interpret them as signs of fortune. This was an inherited practice from Greek and Roman cultures (*Sortes Homericae*—usually from *Iliad*, *sortes Virgilianae*—using *Aeneid* fragments or verses). In France, the Gallican synods of Vannes (465 CE), Agde (506), Orleans (511), and Auxerre (570–590) passed ordinances vowing to excommunicate any Christian who "should be detected in the practice of this art, either as consulting or teaching it'" (Metzger 1993). What is most surprising is that the most well-known instance of *sortes biblicae* was by St. Augustine of Hippo who in the year 386 was prompted by a childlike voice he heard telling him to "take up and read" (in Latin: *tolle, lege*). Augustine opened a Bible at random, selecting from the two sides the verses of Romans 13:13–14 ("Not in rioting and drunkenness, not in chambering and impurities, not in strife and envying; but put you on the Lord Jesus Christ, and make not provision for the flesh in its concupiscences."), and later wrote that "as if before a peaceful light streaming into my heart, all the dark shadows of doubt fled away" (*Confessions*, Bk. 8, Chap. 29). Augustine was then converted, calling the experience a direct work of God, but a few centuries later this would have been considered just a blaspheme and sinful behavior (some millennia later, it could be easily typified as "schizophrenic").

So, Gataker studied lotteries and hazard games because he wanted to clear the darkness inside them, and at the same time to reflect the paucity of the human mind, which always needed the divine omniscient guidance (Rescher 1995).

The next attempt to join theology and statistics was the demographic theology of Johann Peter Süssmilch. In 1741, this German priest with interests in demography published *Die göttliche Ordnung in den Veränderungen des menschlichen Gesglechts, aus der Grut, dem Tode, un der Fortpflanzung* (*The Divine order in the changes in the human sex from birth, death and reproduction of the same*), a very curious work full of still more curious theses, all about the invisible guidance of God through the hand of His Providence. According to Süssmilch, if somebody analyzes long rungs of birth registers, it can be found that approximately the 50 % ratio of males and females is stable. For him, this was a logic consequence of the evident hand of God, and this text becomes one of the first attempts to talk about

---

[12]We will find in one of the leading founders of modern statistics, Laplace, a similar idea: "(probability) is relative, in part to our ignorance, and in part to our knowledge", Laplace (1814: 8).

intelligent design. In fact, Süssmilch had not been the first to point to this fact, but the Scottish John Arbuthnot. In 1710, he published *An argument for Divine Providence, taken from the constant regularity observed in the births of both sexes* in the Royal Society's *Philosophical Transactions*, where he analyzed birth data and demonstrated that males were born at a greater rate than females. He considered that this fact was against the 50 % equal odds and that the only explanation was the active influence of divine providence into this process, in order to correct the early deaths of males who die young more often than females. This problem of sex ratio was attacked but not solved one century later by Charles Darwin and was necessary to reach the twentieth century to find an answer: Ronald A. Fisher established it in 1930 with the book *The Genetical Theory of Natural Selection* the so-called Fischer principle, a ratio of 1:1 between sexes as an evolutionary stable strategy. But cultural incidence is changing this ratio, as has been noted by Hvistendahl (2011).

2.3   Fortuna, Destiny, Luck, Chance, or Probability …

Until this moment we have seen that before seventeenth century, a specific vocabulary to deal with probability did not exist, basically because cultural paradigms cannot allow it inside them. Nevertheless, these different cultures had the necessity to express several notions of non-deterministic events. That is, not controlled events. I will make a short journey across these words and their meanings.

Greek Goddess *Ananké* (*Necessitas* for Roman mythology), the mother of the Moiri and Adrasteia, was considered the Goddess of destiny, necessity, and fate. Supranatural or divine rules guided human lives secretly, who should discover it and embrace their destiny. At a certain level, there was no free will for them just a terrible divinity will. *Tyche* was a different goddess (worshipped in Rome under the name of *Fortuna*) who was considered as the governor of the prosperity or decline of a city as well as the source of all unexpected events in human life, whether good or evil. So in a certain way, she was related to hazard or luck. Temples were built to Tyche asking for a better life and Romans considered her as the *fors*, the luck, fortuity, accident, and chance and sometimes painted her as a woman who spins a wheel, the *Rota Fortunae* (or *wheel of fortune*). Fortuna was also christianized and forms part of the history of medieval art and minds. Curiously, there was still a third goddess, Ananke, also called "Necessity," the strongest force in the realm of gods who was also paired with Fortuna. Something similar exists in Hinduist tradition under the name of *karma* (a cosmic regulatory law of cause–effect, along with *samsara* (reincarnation cycle) *and moksha* (liberation from *samsara*). In our days, the presence in human life of the notion of chance is overwhelmingly present. As the poet tells us, luck is everything.[13] Close to the notion of "fortuna" in the

---

[13]Childish (1988). Poem 'h.m. prison maidstone'. A wonderful poem written by a different poet.

twelfth and thirteenth centuries in Europe appearing in the oriental Mediterranean area, the notion of *risicum* or risk emerged, probably from the Arab word *riszq* (Piron 2004). This *risicum* was related to the games practices as well as to the perils of economic procedures, most of them analyzed by Franciscan monks (Ceccarelli 1999). These theologians were mainly interested in the notion of contract and how random elements present in that contract should be considered (Meusnier and Piron 2007). This was very close to the first maritime insurances invented in Tuscany in the first half of the fourteenth century were the risk of a commercial operation, accepted by the insurer against the payment of a *premium*. For a psychological analysis of luck, see Pritchard and Smith (2004).

Finally, the term "probability" can be found in Classic Rome in Latin as the word "probabilis," translated as "credible," but in 1660 when it turned the meaning toward the modern use (Hacking 1984). It was in 1657 that Hyugens published *De Ratiociniis in Ludo Aleae* (1714 English version published as "The VALUE of all CHANCES IN Games of Fortune; CARDS, DICE, WAGERS, LOTTERIES, &c. Mathematically Demonstrated"). Very soon *La logique, ou l'art de penser*, in 1662 also appeared by Antoine Arnauld and Pierre Nicole, commonly quoted as Port-Royal Logic. They introduced the idea of the necessary quantification of probability.

2.4   *Pay Me Again, Sam...*From New Gods and Taxes to Statistics

Approximately 6/7 years before year 1^{14} of our Era and during the reign of Emperor Augustus, Publius Sulpicius Quirinius was appointed governor of Syria. One of his first actions was to improve his taxes recollection performing a new census of the Jewish population. The *Gospel of Luke* explains that was then when Joseph and the pregnant Maria travelled to Jerusalem to notify their data, but because of the advanced situation of her pregnancy, they gave birth to the child in Bethlehem: he was Jesus, the founder of the biggest and most widespread religion existing to this day. Don't be lost by this shell game with words: here the important thing is the census, not the religion.

If we look at the Oxford English Dictionary, as a simple source to the topic and look for the entry "statistic" this is found:

> The earliest known occurrence of the word seems to be in the title of the satirical work *Microscopium Statisticum*, by 'Helenus Politanus', Frankfort (?), 1672. Here the sense is prob. 'pertaining to statists or to statecraft' (cf. statistical a. 1). The earliest use of the adj. in anything resembling its present meaning is found in mod.L. *statisticum collegium*, said to have been used by Martin Schmeizel (professor at Jena, died 1747) for a course of lectures

---

[14]See Gould (1997) for the debate on when exactly it is supposed that Jesus was born and how it should be considered numerically the first year of Jesus's life.

on the constitutions, resources, and policy of the various States of the world. The G.
statistik was used as a name for this department of knowledge by G. Achenwall in his
Vorbereitung zur Staatswissenschaft (1748); the context shows that he did not regard the
term as novel. The F. statistique n. is cited by Littré from Bachaumont (died 1771); Fr.
writers of the eighteenth century refer to Achenwall as having brought the word into use.
The sense-development of the word may have been influenced by the notion that it was a
direct derivative of L. status.

And by "statistics":

c.B.1.c Statistics. Any of the numerical characteristics of a sample (as opposed to one of the
population from which it is drawn). Cf. parameter 2f.

So, statistics has a direct relationship with *census,* the registration of citizens and
their property for purposes of taxation. From this close tie between social numbers
and government, the etymological trace of the work "statistics" can be understood.
The German jurist and philosopher Gottfried Achenwall coined the word "Statistik"
in his 1752 work *Staatsverfassung der Europäischen Reiche im Grundrisse*
(Constitution of the Present Leading European States), when he related mathematical calculations of country activities like commerce or agriculture. He also gave
currency to the word "Staatswissenschaft" (science of politics), the knowledge
necessary to understand and run a modern State. At the beginning of the section
"Vorbereitung von der Statistik überhaupt," he identifies some authors who in the
past talked about things close to his notion of statistics, a concept that he defined in
section §5: "Staatsverfassung eines oder mehrerer einzelnen Staaten ist die
Statistik" (The constitution of one ore more individual states is the statistics), and in
§6 he added "Durch die Statistik erlangt man die Staatskenntniß" (Thanks to
statistics somebody can achieve knowledge about the State). Achenwall makes a
qualitative approach to numbers and the affairs of the State (in this sense he also
talks of Staatslehre, Staatswissenschaft, Staatrecht...), not merely one quantitative
as we can infer in our days from the notion of "statistics." Going to the core of his
ideas, we find a very charming notion of State, §2: "Staat ist eine Gesellschaft von
Familien, welche zu Beförderung ihrer gemeinsamen Glückseelichkeit unter einem
Oberhaupte mit einander vereiniget leben" (The State is a society of families who
live together under the guidance of a superior power for the conveyance of a
common happiness).

The *Bills of Mortality* (1662, the complete title is *Natural and Political
Observations Made upon the Bills of Mortality*), published by the haberdasher John
Graunt included the first life table and turned his author into one of the first
demographers and epidemiologists. He made statistical analysis of the population of
London and his impressive results appointed him to the election as member of The
Royal Society despite the class reluctances. One of the 12 who were at Gresham
College in November 28,1660, who proposed a new institution that would be the
Royal Society, William Petty is considered, together with Graunt, the founder of the
modern census statistics, basically due to his interests in what he called "political

arithmetic." He made estimations and used simple averages, always as part of his duties working alongside Oliver Cromwell and serving as parliamentarian.

From the "Herodes Census" (Quirinius), to the first colonial census made in Peru by the Spanish Don Pedro de La Gasca at Perú of 1548, the interest in such tables of data was mainly due to economics as well as military. It is not strange that mathematicians, even leading experts like Leibniz, were attracted by governmental forces to this research field, which became secret for national security. At the same time, these huge lists of data required from new ways to be easily understandable, that is, visually mapping. And, finally, with so many objective data, the idea of "normality," emerged that introduced into human studies the notion of "average man" (*home moyen*, according to Quetelet, the astronomer who first applied statistical analyses to human biological domains). This also made possible a mathematization of the whole human sphere, allowing the birth of the social mathematics, by Condorcet, as well as the consequent positivist view of Auguste Comte.

2.5 From Dice to Vaccines and Assurance Companies: The Birth of Probability

1660 is the year, if we follow Hacking (1984, 1990), of the birth of probability. But several things concurred in order to generate the complex and extended notion of probability, which we will analyze in this section.

2.5.1 First of All, Vaccines

The dispute between Daniel Bernoulli and Jean Le Rond D'Alembert on the efficacy and utility of smallpox vaccination was a different context in which the probability issues were discussed, far from previous recreational, hypothetical, or mathematical discussions. It was the year 1760, in the middle of an intense controversy on the benefits of inoculation that had started with the works of Pierre Louis Moreau de Maupertius (1698–1759) and Charles Marie de la Condamine (1701–1774). The latter, especially, has written several memoranda favoring the introduction of inoculation into France, then a very young technique (Dietz and Heesterbeek 2002). Daniel Bernoulli wrote a paper modeling smallpox, using Halley's life table and some data concerning smallpox to show that inoculation was advantageous if the associated risk of dying was less than 11 %. Inoculation could increase life expectancy at birth to up to three years (Bacaër 2011). This was the *first* mathematical model employed in epidemiology, a discipline that we will discuss in later chapters because of its close links with statistics and causality debates. Immediately, D'Alembert criticized Bernoulli's work from a seminal presentation at *Académie royale des sciences* to its several publications.[15] Bernoulli's model was probably the first compartmental model and described the age-specific prevalence of immunes for an endemic infection which is potentially lethal. In a letter to the mathematician Euler, Bernoulli showed himself as sad at D'alembert's criticisms and considered his work as "c'etoit, si j'ose le dire, comme une *nouvelle province* incorporée au corps des mathematiques" (translated as "it was, I dare say, like incorporating a *new province* into the body of mathematics"; cursives are mine). This debate aroused an enthusiasm in France for the social uses of probability (Zabell 2011: 1153). The star-system philosopher Voltaire joined the general debate on probability and its uses incorporated into the list of friendly statistics authors. He wrote a book in 1772 entitles *Essai sur les probabilités en fait de justice*. There he explained (p. 371) that "Presque toute la vie humaine roule sur des probabilité" (Almost all human life is based on "probability"), a curious declaration in a deterministic era, sign of the changes that were happening in his time.

2.5.2 Secondly, Insurance Companies

As a second domain encapsulating an attraction toward the use of numbers to explain and predict future outcomes is the assurances. Yes, a pragmatic use, as usually happens with most new ideas of humanity. Babylonian merchant land traffics and later Phoenician merchant sea traffic were the first situations in which a rude and basic insurance idea was applied. Following Trenerry (2009: 6), the essentials of that Bottomry were reinforced by law for the first time in the Code of Hammurabi (2250 B.C.). Later, Achaemenids (Persians), Greeks, or Romans evolved this simple version and started a transformation (e.g., introducing life insurances by collegia funeraticia in Roman culture[16]) that led to the origin of modern insurance companies in seventeenth century. A first step toward this

---

[15]For the very strange reasons of life, D'Alembert immediately wrote a criticism which he presented on November 12, 1760, to the Royal Academy of Sciences and which he published his collected works in the following year. This means that his critique of Daniel Bernoulli appeared five years before Bernoulli's contribution was eventually published by the Academy in 1766. Bernoulli was very annoyed about the critique by d'Alembert, which can be seen from his letter to Euler in April 1768 (Dietz and Heesterbeek 2002: 12).

[16]A look at an old but still fascinating book like *Die römischen Collegia Funeraticia nach den Inschriften* (1888), by Traugott Schiess, is very informative in this topic. An online version is available at: http://archive.org/stream/diermischencoll00schigoog#page/n5/mode/2up. These societies allowed poor people to cover the expenses of their burial, as well as some other assistance during their life.

process was made in Genoa in 1347, when the first known insurance contract was created (Franklin 2001). The first book on insurances was written in 1557 by the Portuguese lawyer Pedro de Santarem (Petrus Santerna): *Tractatus de assecurationibus et sponsionibus mercatorum ad praxim quotidianam utilissimus & omnibus in foro presertim mercatorum versantibus quotidianus*. In this book, sea trade protection (*Mercatores maris*) was the basic target of the insurance system protecting merchandise as well as the crew or the vessel. The creation of quantitative lists trying to evaluate the possible risks from the destiny, boat, captain, crew, or merchandise required from mathematical tools from statistical nature. A coffee house led by Edward Lloyd in London in 1688 was a common place for sailors and shipping industry investors to share last notices about the field, and it was a source of information for insurance experts. It was in this conceptual arena in which ideas like "normal curves" or "normal man" (Adolphe Quetelet's l'homme moyen) emerged from statistical data to enter into political, medical, artistic, and even the anthropological arena. Fire protection was another important speciality of insurance companies, and curiously, Benjamin Franklin founded in 1752 America's oldest, continuously active insurance company: *Philadelphia Contributorship for the Insurance of Houses from Loss by Fire*. The Contributorship, as is now its common reference, was a proactive insurance carrier refusing to provide coverage to houses and other structures that were not constructed according to strict building standards.

2.5.3 Third, Legal Issues and the Notion of "Evidence"

In Sect. 2.5.1, it was mentioned that Voltaire wrote a book on probability and evidence in the legal context. This work inspired Minister Turgot in the reform of the French legal system and prepared the field for fruitful research across the time. In 1837, for example, the mathematician Siméon-Denis Poisson wrote *Recherches sur la probabilité des judgements*, where he made an interesting distinction between subjective and objective senses of probability (Zabell 2011: 1153). At a certain level, he followed some previous but not well-defined concepts of Hume. The British philosopher wrote in 1739, *A Treatise on Human Nature* (T 1.3.11.3, SBN 124–125): "Probability or reasoning from conjecture may be divided into two kinds, viz. that which is founded on *chance*, and that which arises from *causes*. We shall consider each of these in order." For Hume, chance was merely the negation of a cause and causes themselves were not real just mental habits, and even more "chance is nothing real in itself" (*ibid.* p. 125). Without causality, all the beliefs of Enlightenment religious scientists, who looked at nature to find the justification of the existence of God and his rules, disappeared. Then, the study of causal events became a priority for theologians. Probability was then the result of an imperfect experience[17] and consequently chances were equal and indifferent (no place for statistical multicausality). Poisson tried to make a new objective field of research, avoiding previous errors and misunderstandings or fatidic double senses, basically those concerning the meaning of words, into a philosophical problem: "Fundamentally, the theory of chance and mathematical probability applies to two kinds of questions that are quite distinct: to questions of *possibility* which have an objective existence, as has been explained, and to questions of *probability* which are relative, in part to our knowledge, in part to our ignorance" (quoted from Zabell 2011: 1156). This philosophical debate continued with Cournot in his 1843 *Exposition* and 1851 *Essai,* where he affirmed the existence of two kinds of probability (and types of studies): philosophical and mathematical. First one was not reducible to a calculus of chance, while the second was. Philosophical probability was closer to natural phenomena, and out of the realm of mathematical analysis, and hence, of absolute (frequentist) truth. This distinction between *chance* (philosophical) and *probability* (mathematical) was also proposed in a certain way by Jakob Fries in Germany (1842, *Versuch einer Kritik der Prinzipien der Wahrscheinlichkeitsrechnung,* Braunschweig: Vieweg),[18] Richard Leslie Ellis (1843, "On the Foundations of the Theory of Probabilities", *Transactions of the Cambridge Philosophical Society* vol 8) and John Stuart Mill (1843, *A System of Logic, Ratiocinative and Inductive*) in England. It is interesting to note that Mill was hostile to statistical thinking and related thinking (like Quetelet's social physics). Auguste Comte, another influential thinker of that century, was an opponent of social statistics as well as of other, for him, biased uses of statistics into scientific realms, despite his defense of scientific quantification. In the case of Mill, his approach to the statistical debate came from the analysis of coincidences: how to distinguish coincidences that are casual from those that come from natural laws or processes. Mill argued against Laplace's (initial) subjective interpretation of probability and made him affirm that statistics misuses were "the real opprobrium of

---

[17]Specifically he wrote (*ibid.* T 1.3.12.25, SBN 142): "BUT beside these two species of probability, which are deriv'd from an *imperfect* experience and from *contrary* causes, there is a third arising from ANALOGY, which differs from them in some material circumstances. According to the hypothesis above explain'd all kinds of reasoning from causes or effects are founded on two particulars, viz. the constant conjunction of any two objects in all past experience, and the resemblance of a present object to any one of them. The effect of these two particulars is, that the present object invigorates and in livens the imagination; and the resemblance, along with the constant union, conveys this force and vivacity to the related idea; which we are therefore said to believe, or assent to." The complete work of Hume, very professionally digitalized, can be accessed from http://www.davidhume.org.

[18]Fries, besides of his studies on Kantian psychology and ethics, wrote in 1816 a text *How the Welfare and Character of the Germans are Endangered by the Jews* (Über die Gefährdung des Wohlstandes und Charakters der Deutschen durch die Juden. Eine aus den Heidelberger Jahrbüchern der Litteratur besonders abgedruckte Recension der Schrift des Professors Rühs in Berlin: Ueber die Ansprüche der Juden an das deutsche Bürgerrecht, Heidelberg: Mohr und Winter). It is very curious that statistics was so close related (or used) in a future with eugenics and racist theories, as we will see with the use of Darwin's work by Galton. *L'esprit du siècle* was a bad spirit...directing European culture to the worst and darkest future.

mathematics." (ibid. p. 382).[19] Under this accusation, we must identify a debate on the procurance of inferences. One of the theoreticians involved in this debate was Charles S. Peirce, who wrote on the topic on *Illustrations of the Logic of Science* (1877–1878) and *A Theory of Probable Inference* (1883). He conducted research on regression models and defended a propensity theory of probability (later continued by eminent philosopher Karl Popper together with his ideas on falsifiability, close to the null hypothesis testing ideas, as suggested by Meehl in 1967, hinting at achieving the Popperian principle of representing theories as null hypotheses and subjecting them to challenge).

Close to the Ellis probability ideas, in 1866 the British Philosopher and logician John Venn wrote *The Logic of Chance: An Essay on the Foundations and Province of the Theory of Probability*, which is considered one of the first texts on frequentist statistics paradigm, something we have still not explained and that we will analyze in the following chapters. Venn was also the inventor of the very important *Venn diagrams*, and one of these can be seen as peacefully mingling colors from the sun light at a stained glass window in the dining hall of Gonville and Caius College, in Cambridge (UK). In the preface to the first edition, Venn explained:

> This supposed want of harmony between Probability and other branches of Philosophy is perfectly erroneous. It arises from the belief that Probability is a branch of mathematics trying to intrude itself on to ground which does not altogether belong to it. I shall endeavour to show that this belief is unfounded. To answer correctly the sort of questions to which the science introduces us does generally demand some knowledge of mathematics, often a great knowledge, but the discussion of the fundamental principles on which the rules are based does not necessarily require any such qualification. (...) The opinion that Probability, instead of being a branch of the general science of evidence which happens to make much use of mathematics, is a portion of mathematics, erroneous as it is, has yet been very disadvantageous to the science in several ways. Students of Philosophy in general have thence conceived a prejudice against Probability, which has for the most part deterred them from examining it.

Venn considered that the foundations of probability needed to be explained as well as the same probability. That mathematicians were not interested in entering into this philosophical quicksand was not an excuse about the necessity of the project. The philosophical debate surrounding the statistical analysis of nature has disappointed, hassled, abashed, or disgusted those researchers with a more mathematical training and pragmatic spirit, although at the end the philosophical debate appeared again to justify the "best" statistical approach. The debate on statistics is not only

---

[19]The exact and full quote is as follows: "It is obvious, too, that even when the probabilities are derived from observation and experiment, a very slight improvement in the data, by better observations, or by taking into fuller consideration the special circumstances of the case, is of more use than the most elaborate application of the calculus to probabilities founded on the data in their previous state of inferiority. The neglect of this obvious reflection has given rise to misapplications of the calculus of probabilities which have made it the real opprobrium of mathematics. It is sufficient to refer to the applications made of it to the credibility of witnesses, and to the correctness of the verdicts of juries". To provide justice to his words, Mill did not have a negative attitude towards statistics, just to some (for him) misuses of statistical tools.

ontological, but also epistemological: the idea of method and truth. Venn even included some exceptions to this philosophical disinterest, like de Morgan's *Formal Logic* and Boole's *Laws of Thought*. He also considered Mill and Whewell (1840, *Philosophy of the Inductive Sciences*) as trend inducers to an aversion toward interest on probability, something wrong and misleading. Herschel, a leading thinker at that time, considered probability in his *preliminary Discourse* (1830) only as something related to measurement techniques, but not as a basic characteristic of normal science.

Finally, the introduction of statistics into the judicial arena was seen as a great mistake for several authors although Joseph Louis François Bertrand, for example, wrote against this common view in his 1889 *Calcul des probabilités*, p. 43, 5: "L'application du calcul aux decisions judiciaires est, dit Stuart Mill, le scandale des Mathematiques.[20] L'accusation est injuste. On peut peser du cuivre et le donner pour or, la balance reste sans reproche. Dans leurs travaux sur la theorie des jugements, Condorcet, Laplace et Poisson n'ont pese que du cuivre."

After all these seminal ideas on probability, we are now prepared to introduce the reader to Prof. Bayes and his revolutionary ideas on numbers and events.

References

Bacaër, N. (2011). Daniel Bernoulli, d'Alembert and the inoculation of smallpox (1760). In N. Bacaër (Ed.), *A short history of mathematical population dynamics* (pp. 21–30). UK: Springer.

Bellhouse, D. R. (1988). Probability in the sixteenth and seventeenth centuries: An analysis of puritan casuistry. *International Statistical Review, 56*(1), 63–74.

Burton, D. (2005). *The history of mathematics: An introduction.* USA: McGraw-Hill.

Ceccarelli, G. (1999). Le jeu comme contrat et le *risicum* chez olivi. *Études de philosophie médiévale, 79,* 2–25.

Childish, B. (1988). *To the quick.* UK: Hangman.

Cournot, A.A. (1843) Exposition de la théorie des chances et des probabilités. In B. Bru (Ed.) (19984) *Œuvres complètes.* Tome I, Paris: Vrin.

Crosby, A. W. (1996). *The measure of reality: Quantification in Western Europe, 1250–1600.* UK: Cambridge University Press.

Dandoy, J. R. (2006). Astragali through time. In M. Maltby (Ed.), *Integrating zooarchaeology* (pp. 131–137). Oxford: Oxbow Books.

---

[20]Bertrand refers to Mill's notion of "the real opprobrium of mathematics", in his *A System of Logic Ratiocinative and Inductive., Chapter 18. Of The Calculation Of Chances.* The full section says" It is obvious, too, that even when the probabilities are derived from observation and experiment, a very slight improvement in the data, by better observations, or by taking into fuller consideration the special circumstances of the case, is of more use than the most elaborate application of the calculus to probabilities founded on the data in their previous state of inferiority. The neglect of this obvious reflection has given rise to misapplications of the calculus of probabilities which have made it *the real opprobrium of mathematics*. It is sufficient to refer to the applications made of it to the credibility of witnesses, and to the correctness of the verdicts of juries."

David, F. N. (1998). *Games, gods and gambling, a history of probability and statistical ideas.* Mineola: Dover.

Devlin, K. (2008). *The unfinished game: Pascal, fermat, and the seventeenth-century letter that made the world modern.* USA: Basic Books.

Diels, H. (1903). *Die Fragmente der Vorsokratiker* (p. 55). Berlin: Weidmannsche Buchhandlung.

Dietz, K., & Heesterbeek, J. A. P. (2002). Daniel Bernoulli's epidemiological model revisited. *Mathematical Biosciences, 180,* 1–21.

Edmunds, L. (1972). Necessity, chance, and freedom in the early atomists. *Phoenix, 26,* 342–357.

Franklin, J. (2001). *The science of conjecture: Evidence and probability before pascal.* Baltimore: Johns Hopkins University Press.

Gabbay D. M., Thagard, P., & Woods, J. (Eds.) (2011) *Handbook of the Philosophy Volume 7, Philosophy of Statistics.* Oxford (UK): Elsevier.

Gigerenzer, Gerd, et al. (1989). *The empire of chance: How probability changed science and everyday life.* UK: Cambridge University Press.

Gould, S. J. (1997). *Questioning the Millennium.* USA: Harmony Books.

Hacking, Ian. (1984). *The emergence of probability: A philosophical study of early ideas about probability, induction and statistical inference.* UK: CUP.

Hacking, Ian. (1990). *The taming of chance.* UK: Cambridge University Press.

Hald, A. (1988). *A history of probability and statistics and their applications before 1750.* NY: Wiley.

Holmes, A. (1826). *The annals of America from the discovery by Columbus in the year 1492* (p. 142). Cambridge (MA): Hilliard & Brown.

Hvistendahl, M. (2011). *Unnatural selection: Choosing boys over girls and the consequences of a world full of men.* USA: Public Affairs.

Metzger, B. M., & Coogan, M. D. (Eds.). (1993). *The oxford companion to the bible.* Oxford: OUP.

Meusnier, N., & Piron, S. (2007). Medieval probabilities: A reappraisal. *Journal Electronique d'Histoire des Probabilitiés et de la Stadistique, 3*(1), 1–5.

Moussy, C. (2005). Probare, probatio, probabilis dans de vocabulaire de la démonstration. *Pallas, 69,* 31–42.

Northbrooke, J. (1577). *Spiritus est vicarious Christi in terra: A treatise wherein dicing, dauncing, etc. are reproved.* London: Bynneman.

Piron, S. (2004). L'apparition du *resicum* en Méditerranée occidentale, XIIe- XIIIe siècles. In E. Collas-Heddeland, et al. (Eds.), *Pour une histoire culturelle du risqué.* Editions Histoire et Anthropologie: Strasbourg.

Pritchard, D., & Smith, M. (2004). The psychology and philosophy of luck. *New Ideas in Psychology, 22*(1), 1–28.

Rescher, N. (1995). *Luck, the brilliant randomness of everyday life.* NY: Farrar Straus.

Trenerry, C. F. (2009). *The origin and early history of insurance: Including the contract of bottomry.* Clarck (NJ): The Lawbook Exchange Ltd.

Vallverdú, J. (2011a). Bayesian versus frequentist statistical reasoning. In M. Lovric (Ed.) *International Encyclopedia of Statistical Science* (pp. 133–135).

Vallverdú, J. (2011b). History of probability. In M. Lovric (Ed.) *International Encyclopedia of Statistical Science* (pp. 1126–1128).

Zabell, S. L. (2011). The subjective and the objective. In D. M. Gabbay, P. Thagard, & J. Woods (Eds.) *Handbook of the Philosophy Volume 7. Philosophy of Statistics* (pp. 1149–1174). Elsevier: Oxford (UK).

Chapter 3
The Bayesian Approach and Its Evolution Until the Beginning of the Twentieth Century

Abstract Rev. Bayes and his friend Richard Price created a new way to deal with the philosophical and theological problems on induction as were explained by Hume. This mathematical formula included the notion of subjective probabilities and, consequently, opened a debate on its validity. French mathematician Pierre-Simon Laplace applied it successfully to astronomical calculations just before starting to change his mind over the correctness of Bayes' formula. Several objections and practical challenges made the general implementation of Bayes' ideas impossible.

Keywords Bayes · Bayes' theorem · Price · Laplace · Priors · Subjective probabilities · Induction · Hume · Military

The Presbyterian minister Thomas Bayer died in the serenely beautiful city of Tunbridge Wells, which is 35 miles southeast of London, in April 7, 1761. He had devoted his life to his ministry duties but always keep in mind a true interest in mathematics. Although he published two books[1] his greatest contribution to science remained hidden in his personal writings. Bayes' family asked Richard Price, an Unitarian minister and friend of Bayes,[2] to examine his unpublished papers and it was then that Price realized their importance: One of the manuscripts, *An Essay towards solving a Problem in the Doctrine of Chances*, was extremely important and consequently was read by Price to the Royal Society in 1763. For that occasion,

---

[1] *Divine Benevolence, or an Attempt to Prove That the Principal End of the Divine Providence and Government is the Happiness of His Creatures* (1731) and *An Introduction to the Doctrine of Fluxions, and a Defence of the Mathematicians Against the Objections of the Author of the Analyst* (published anonymously in 1736).

[2] Price was even a beneficiary of Bayes' inheritance. Bayes left £200 to be divided between John Hoyle and Richard Price (Dale 1999: 26). Hoyle was the minister at Stoke Newington from 1748 to 1756. When Hoyle left Stoke Newington to take up a position in Norwich (Browne 1877), Richard Price became the pastor at Stoke Newington. Both chapels eventually became Unitarian churches, and both Hoyle and Price were known Arians, (Bellhouse 2004: 11). It was a complete religious environment.

© The Author(s) 2016
J. Vallverdú, *Bayesians Versus Frequentists*,
SpringerBriefs in Statistics, DOI 10.1007/978-3-662-48638-2_3

Price wrote a very interesting introduction[3] in which he presented Bayes' results as experimental philosophy, the cornerstone idea of English philosophers of that epoch.[4] The philosophical implications of the manuscript were clear: in order to give "a clear account of the strength of analogical or inductive reasoning," according to Price's own words, Bayes, again according to the interpretation of Price, had suggested firmly[5]:

> to find out a method by which we might judge concerning the probability that an event has to happen, in given circumstances, upon supposition that we know nothing concerning it but that, under the same circumstances, it has happened a certain number of times, and failed a certain other number of times. He adds, that he soon perceived that it would not be very difficult to do this, provided some rule could be found, according to which we ought to estimate the chance that the probability for the happening of an event perfectly unknown, should lie between any two named degrees of probability, antecedently to any experiments made about it; and that it appeared to him that the rule must be to suppose the chance the same that it should lie between any two equidifferent degrees; which, if it were allowed, all the rest might be easily calculated in the common method of proceeding in the doctrine of chances.

At any rate, and more according to his true vital interests, after discussing de Moivre's work, Price stated:

> The purpose I mean is, to shew what reason we have for believing that there are in the constitution of things fixt laws according to which events happen, and that, therefore, the frame of the world must be the effect of wisdom and power of an intelligent cause; and thus to confirm the argument taken from final causes for the existence of the Deity.

What motivated Price to work on this paper was that to him the result provided a proof of the existence of God, as well as contributing to clarify the theological and philosophical corollaries of Hume's critics on induction (Bellhouse 2004: 24). When the essay was published posthumously in the *Philosophical Transactions of the Royal Society of London* (1763), it received some good reviews, but it did not receive the surely deserved honors...it was not until the 1950s that it would be applied extensively by different authors. Why? The reasons are clear: because of the calculations complexity of its implementation into real scientific studies, Bayes' theorem was not an easy choice and consequently for two centuries, it remained almost hidden.

---

[3]Although Price himself expressed his role as simple transmitter of the ideas of Bayes, some reluctance could be expressed about his complete respect of the original manuscript. Following some remarks of Prof. Bellhouse (2002), it might be very probable that Mr. Price made some "adjustments" to the final redaction, although he respected the main ideas (see Earman (1992) for a technical analysis).

[4]Following Gillies (1987, as well as by personal correspondence), we must say that Bayes made the mathematical contributions and Price the philosophical arrangement of it. Though the papers do not explicitly cite Hume, there is evidence that the authors were trying to solve Hume's problems about induction.

[5]The original essay can be freely downloaded as a PDF file from http://www.stat.ucla.edu/history/essay.pdf.

But which was the *exact* contribution of Rev. Bayes? When we read the original paper published posthumously by Price,[6] we find the statement of a problem, how to find for an unknown event the chance that the probability of its happening in a single trial, and several sections with a conceptual response to the initial problem. In the same problem description, Bayes affirms that this chance lies somewhere between any two degrees of probability that can be named. Then, he elaborated a list of definitions, trying to clarify frequent linguistic misunderstandings when we talk about chance, and 10 propositions, 3 rules plus 1 appendix later, he closed his text affirming that:

> But what most of all recommends the solution in this *Essay* is, that it is compleat in those cases where information is most wanted, and where Mr. De Moivre's solution of the inverse problem can give little of no direction; I mean, in all cases where either p or q are of no considerable magnitude. In other cases, or when p and q are very considerable, it is not difficult to perceive the true of what has been here demonstrated, or that there is reason to believe in general that the chances for the happening of an event are to the chances for its failure in the same *ratio* with that of p to q. But we shall be greatly deceived if we judge in this manner when either p or q are small. And tho' in such cases the *Data* are not sufficient to discover the exact probability of an event. Yet it is very agreeable to be able to find the limits between which it is reasonable to think it must lie, and also to be able to determine the precise degree of assent which is due to any conclusions or assertions relating to them.

In any case, here there is no trace of any formula that synthesizes his ideas! We must go to France to find a very young mathematician with the answer. The main ideas in Bayes' theorem were rediscovered independently by Pierre Simon Laplace,[7] who first published his version in 1774, eleven years after Bayes, in one of Laplace's first major works when we was only 25 years old: "*Mémoire sur la probabilité des causes par les évènements.*" Although he made the earliest application of "Bayesian" ideas to multimodal setting (Stigler 1986: 361), the principal success of Laplace was the spreading of these new ideas (from Bayes-Price-Laplace) to the whole mathematical community. In fact, Bayes own article was ignored until 1780 and had no important role in scientific debate until the second half of the twentieth century.

Perhaps, it was in 1781, when Richard Price visited Paris, when Bayesian words reached Laplace for the first time reinforcing his preliminary ideas. But Laplace made something even more important: he created the formula of the (not completely honestly) so-called Bayes' theorem. It was in 1812, in his book *Théorie analytique des probabilités* (TAP), specifically into section II, §1 (and with some slight changes into the 1814 second edition), where he presented the conceptual basis of the formula:

> La probabilité d'un événement future, tirée d'un événement observé, est le quotient de la division de la probabilité de l'événement composé de ces deux événements, et determiné a priori, par la probabilité de l'événement observé, déterminée parcillement a priori.

---

[6]Downloadable from the Royal Society at http://rstl.royalsocietypublishing.org/content/53/370.

[7]There is a different possibility that Laplace knew about the Bayes–Price ideas through a common friend, Condorcet. Unfortunately, this is only a possibility that cannot be confirmed empirically.

Hald (2007) considers that Laplace's principle could have two interpretations: (a) a frequency interpretation (based on a two-stage model with objective probabilities and (b) an interpretation based on the principle of insufficient reason (also called the *principle of indifference*). Replacing the old and original terminology employed by Laplace, the proof acquires the modern form as follows:

$$P(C_i|E) = \frac{P(C_i)P(E|C_i)}{\sum P(C_i)P(E|C_i)}$$

where $i = 1,..., n$.

Despite the two possible interpretations about the real Laplacian contribution, as a frequency interpretation based on a two-stage model with objective probabilities and an interpretation based on the principle of insufficient reason, it is obvious that he was the person who transformed the general ideas of Bayes into a well-defined theorem. Nevertheless, for Laplace, probability was the result of our ignorance not an ontological and epistemological reality of the world. As Laplace wrote in TAP, iv: "La courbe décrite par une simple molécule d'air ou de vapeurs, est réglée d'une manière aussi certaine, que les orbites" planétaires; il n'y a de différence entre elles, que celle qu'y met notre ignorance. La probabilité est relative en partie à cette ignorance, et en partie à nos connaissances." And the interest about probability is at the same time not a true scientific project, but most of the time it is the result of our fear about future (TAP, 2nd edition, *xv*: "La probabilité des événemens sert à déterminer l'espérance ou la crainte des personnes intéressées à leur existence").

It can be affirmed that Laplace reached the same basic ideas of Bayes independently, but as soon as in the year of 1814, with the second edition of TAP, page *ciii*, Laplace cited for the first time the ideas of Bayes: "Bayes, dans les Transactions Philosophiques de l'année 1763, a cherché directement la probabilité que les possibilités indiquées par des expériences déjà faites, sont comprises dans les limites données; et il y est parvenu d'une manière fine et très-ingénieuse, quoiqu'un peu embarrassée. Cet objet se rattache à la théorie de la probabilité des causes et des événemens futurs, conclue des événemens observés; théorie dont j'exposai quelques années après, les principes, avec la remarque de l'influence des inégalités qui peuvent exister entre des chances que l'on suppose égales. Quoique l'on ignore quels sont les événements simples que ces inégalités favorisent; cependant cette ignorance même accroit souvent, la probabilité des événemens composés." He recognized the genuine ideas of Bayes but at the same time affirmed that "he (for Bayes) did it in a way very fine and ingenious, though somewhat awkwardly."

The case study that Laplace chose in order to check his formula was gender birth records differences. The largest data set available for him was the census data, and at the same time, there was an unclosed debate about one related question: It is true that there are slightly born more boys than girls, but... is this a constant phenomenon of Nature or instead of it, just an accidental and provisional anomaly? From a pure statistical point of view, both sexes should be born with the same frequency of 50 % odds. Using recorded data from Paris, London, Naples, and other cities, we stated in 1812 that it was a general law for the human race. His methods were applied by his friend Alexis Bouvard to the astronomical calculus of the masses of Jupiter and Saturn, with such a high accuracy that Laplace understood the power of the statistical tools until the point to affirm the superiority of this new science (classic + statistical) over any other discipline to explain the world.[8] A short anecdote explains it perfectly: in 1796 Emperor Napoleon I received Laplace after his publication of the *Exposition du système du monde* (*The System of the World*). Once there, Emperor asked Laplace "Newton a parlé de Dieu dans son livre. J'ai déjà parcouru le vôtre et je n'y ai pas trouvé ce nom une seule fois." (Newton spoke of God in his book. I've looked at yours but I did not find his name even once), and he replied "Citoyen premier Consul, je n'ai pas eu besoin de cette hypothèse." (Citizen First Consul, I have not necessity of that hypothesis."), (Fayé 1884, 109– 111).[9] Going beyond Newton,[10] Laplace choose a new direction for science, far from supranaturalistic forces, but still inside a deterministic world, where statistics was at the end, the tool to solve our lack of deep understanding about the world.

Also inside TAP, we can find one of the greatest contributions of Laplace: the central limit theorem. According to Fischer (2010: 18), Laplace's ideas were applied to two categories: "sums of random variables" and "inverse probabilities." Initially, by 1774, his methods involved only a posteriori probabilities, but from 1810, he was able to adapt his method to a priori probabilities,[11] making approx- imations of probabilities of sums of independent random variables (something that can be considered the first step toward the creation of the central limit theorem).

---

[8]Dealing with large amounts of complex data was a problem that emerged at the end of the eighteenth century and became a true problem for nineteenth-century scientists. New evidences and calculations in Astronomy, huge amounts of natural collected biological data required a fundamentally new way of thinking. It was provided by statistical thinking. Was the universe stable thanks to a Newtonian God? Besides the lack of knowledge about the real mechanistical cause of gravitatory, then still a dark force (the Higgs boson that justified gravity as undiscovered until 2013, at the European LHC), the gravitatori calculations were so complex that their calcu- lation power was not able then to do it. As a consequence, statistical approximations were the only rational solution to that problem, and it's what Laplace understood (Bertsch McGrayne 2011: 19).

[9]This attitude can be labeled as "infidel mathematics" and was run by the free thinkers of the Enlightenment. On the other side, religious reformists that were considered Dissenters were interested in using mathematics to do the opposite: use mathematics to justify the existence of God (Bertsch McGrayne 2011: 3–5).

[10]Newton, opposed Laplace in this point, claimed against hypotheses: just remember his "hy- potheses non fingo," but at the same time was not able to explain gravity and the necessity of a God as a Watchmaker inside a clock-universe. Clearly, eighteenth-century rationalism is not a unified intellectual project, nor a true naturalist rationalism. Anyhow, Laplace was accused by radical revolutionaries as "Newtonian idolator" and he was arrested on suspicion of disloyalty to the French Revolution. Science under politics debates. Some centuries later, the Lysenko case repeated this situation under the communism of the URSS.

[11]Even in that case only to *uniform priors* following the insightful suggestions of Stigler (2012, from personal electronic epistolary talk). As he wisely points out, before Galton there was no real multivariate analysis, so the modern practice of starting with a prior and likelihood then getting the multivariate and then the conditional (posterior) distribution could not be done before the 1880s.

These ideas deeply influenced the probability and error theory during the nineteenth
century, under the basic idea that probability was "good sense reduced to calculus"
(TAP, cv: "On voit par cet Essai, que la théorie des probabilités n'est au fond, que
le bon sens réduit au calcul"). Curiously, in the last decade of his life, Laplace
turned himself into a 'frequentist' statistician... he, who was the first Bayesian!
This mental shift toward a more strict view on causality led Laplace to become a
determinist, following in some ways the ideas of Leibniz. The whole book *Essai
philosophique sur les probabilités*, 1814, is the work in which he defends his
change:

> Nous devons donc envisager l'état présent de l'universe comme l'effet de son état antérieur,
> et comme la cause de celui qui va suivre. Une intelligence qui pour un instant donné
> connaîtrait toutes les forces dont la nature est animée et la situation respective des êtres qui
> la composent, si d'ailleurs elle était assez vaste pour soumettre ces données à l'analyse,
> embrasserait dans la même formule les mouvements des plus grands corps de l'universe et
> ceux du plus léger atome; rien ne serait incertain pour elle, et l'avenir comme le passé serait
> présent a ses yeux.[12]

This man with deep knowledge of all possible states of the universe was called later
Laplace's demon, or even sardonically by Reichenbach (1952: 226) *Laplace's
superman*. Here, Reichenbach demonstrated that Boltzmann's interpretation of the
second law of thermodynamics was confirmed after Heisenberg's principle of
indeterminacy: The vagueness found at quantum mechanical phenomena is nothing
to do with imperfections of the human observer, but is grounded in the physical
structure of the world. So, twentieth century physics diluted any attempt to belief in
the perfect notion or deterministic causation as well as introducing the idea of
meta-level behavior: The world seems to show different behaviors at different
scales.[13]

In any case, and far from the solution of the debate about the true inventors of
the Bayes' theorem, Bayes or Laplace, I will proceed to explain it in its modern
form. Let me introduce you to details of Bayes' theorem and its context.

(a) *Bayes' theorem context.* In eighteenth-century England, there was a debate
    that had ancient roots but at the same time was formulated under a new
    approach: causality and inverse probability. Let us to see them separately:

---

[12]Translated as "We may regard the present state of the universe as the effect of its past and the
cause of its future. An intellect which at a certain moment would know all forces that set nature in
motion, and all positions of all items of which nature is composed, if this intellect were also vast
enough to submit these data to analysis, it would embrace in a single formula the movements of the
greatest bodies of the universe and those of the tiniest atom; for such an intellect nothing would be
uncertain and the future just like the past would be present before its eyes."

[13]There is also the contrary question of scale invariance and dilation, but this is a different question
to be analyzed in a different place. We have neither space here to analyze the problem of con-
necting the gap between quantum microphysics with macrophysics (through mesophysics).
Quantum paradoxes are also a crucial challenge for contemporary thinkers and scientists. For an
attempt to unify quantum microphysics with macrophysics see Sewell (2002).

Causality has been at the core of any philosophical attempt to understand or explain the world, from Western as well as eastern traditions. Perhaps, the term "causality" in itself has not been the exact word we can find in all these ancient books, although the notion of direct sequentiality among events is always present, even in the case of considering it as an illusion of mind and, therefore, to deny it. In Sect. 2.5.3, we already talked about this notion as well as the impact of the ideas of Hume.

Inverse Probability is the second important concept to be explained. The notion of "inverse probability" was first used in English by Augustus de Morgan in the 1830s (Dale 1999: 4) abandoning the notion of "the converse problem" defended by Price for that of "the inverse method." But...what are we talking about? Sometimes we need to infer some information about why something has happened in order to explain a possible similar outcome in the future. These hidden causes can be considered as variables that are still unobserved but at the same time must show some probability distribution, even from an abstract perspective. Obviously, this is an inverse or go-back (causal) problem, and when we introduce a probabilistic quantification of those hidden variables, then we are working with inverse probability.

Bayes tried to design a method that could find causality among events until then considered outside of possible knowledge but at the same time not only applied to future events but also to past ones. In fact, the knowledge of past events is very useful in order to understand the causal chain that led to the present and, henceforth, will shape the future. His solution was presented in his book *An Essay towards solving a Problem in the Doctrine of Chances* (1763). But was Laplace, who gave an exact formulation of the inverse probability in his *Sur la probabilité des causes* (1774), in a scientific context? In modern terms, his contribution was to consider an event E which could be produced by any one of a number of mutually exclusive and exhaustive cases C_i, each of positive probability, and then, for each i:

$$\Pr[C_i|E] = \Pr[E|C_i]\Pr[C_i]/\sum_j \Pr[E|C_j]\Pr[C_j].$$

(b) ***Bayes' theorem with more* detail.**

The modern formulation of Bayes' theorem or Bayes' rule (Vallverdú 2011a) can be commonly found as follows:

$$P(A|B) = \frac{P(B|A)P(A)}{P(B)}.$$

where

- $P(A|B)$ is the *conditional probability* of A, given B. It is also called the *posterior probability* because it is derived from or depends upon the specified value of B.

- $P(B|A)$ is the conditional probability of B given A.
- $P(A)$ is the *prior probability* or *marginal probability* of A. It is "prior" in the sense that it does not take into account any information about B.
- $P(B)$ is the prior or marginal probability of B and acts as a *normalizing constant*.

We can see, then, that our posterior belief $P(A|B)$ is calculated by multiplying our prior belief $P(A)$ by the likelihood $P(B|A)$ that B will occur if A is true. Although Bayes' method was enthusiastically taken up by Laplace and other leading probabilists of the day, it fell into disrepute in the nineteenth century because they did not yet know how to handle prior probabilities properly. The prior probability of A represents our best estimate of the probability of the fact we are considering prior to attending to a new piece of evidence. Therefore, in the Bayesian paradigm, current knowledge about model parameters is expressed by placing a probability distribution on the parameters, called the "prior distribution." When new data become available, the information they contain regarding the model parameters is expressed in the "likelihood," which is proportional to the distribution of the observed data given the model parameters. This information is then combined with the prior to produce an updated probability distribution called the "posterior distribution," on which all Bayesian inference is based. So, Bayes' theorem, an elementary identity in probability theory, states how the update is done mathematically: The posterior is proportional to the prior times the likelihood.

Bernardo (2011: 263) defends the objectivity of Bayesian methods.

> Bayesian methods may be derived from an axiomatic system and provide a *coherent* methodology which makes it possible to incorporate relevant initial information, and which solves many of the difficulties which frequentist methods are known to face. (...) This leads to *objective* Bayesian methods, objective in the precise sense that their results, like frequentist results, only depend on the assumed model and the data obtained. The Bayesian paradigm is based on an interpretation of probability as a *rational conditional measure of uncertainty*, which closely matches the sense of the word 'probability' in ordinary language.[14]

This led to some authors who name this approach as "probability theory." After the seminal work of Bayes and Laplace, we will find a leading defender of it in Harold Jeffreys. He read Karl Pearson's *Grammar of Science* in 1914, and it made a great impression on him, especially the notion of the probabilistic basic of scientific inference. But for Jeffreys, probability was a degree of a reasonable belief, not an objective property of the world. This was an idea shared by some Cambridge thinkers like W.E. Johnson, J.M. Keynes, and C.D. Broad. Following an epistemological approach to statistics in science, Jeffreys asked himself the use of

---

[14]According to Efron (2012: 133), *Objective Bayes* is the contemporary name for Bayesian analysis carried out in the Laplace–Jeffreys manner. According to Gillies (2000), Chap. 3, the interpretation of probability as degree of rational belief by Objective Bayesians has never been resolved and has even given rise to several paradoxes. But as far as I can see paradoxes never stopped scientists or theory-users in their activities. Conceptual paradigms are very flexible as well as resistant to internal or external criticisms.

statistics in scientific practices, something that recently Galavotti (2003) called the "Harold Jeffreys' Probabilistic Epistemology." In 1939, he wrote an influential book *Theory of Probability* published by Oxford University Press, and by probability, he referred to a theory of inductive inference founded on the principle of inverse probability. He maintained some exciting (and even disgusting) debates with R.A. Fisher, which we will analyze in Chap. 5. It can be affirmed that modern Bayesianism was founded by Jeffreys, and his book offered a solid way to introduce Bayesian tools into scientific domains. The main problem is that he was not as persuasive as Fisher, and at the same time, his methods were much more complicated than those of Fisher. Finally, quantum mechanics, then a hot and attracting field, employed frequentist tools rather than Bayesian, and all these factors (including the fierce opposition of Fisher and Neyman-Pearson to Bayesianism, beyond their own confrontation within the "frequentist paradigm") forced Bayesianism into oblivion.

Despite the initial theoretical debates, some practical results gained credit for Bayesian methods. French military[15] were involved in this process thanks to the work of Joseph Louis François Bertrand and his contributions for artillery, a disciplined faced with a host of uncertainties (enemy's precise location, air density, wind direction, variations among hand-forged cannons, and range/direction/initial speed of projectiles). Bertrand made a variation of Bayesian ideas (rejecting universal 50–50 odds for prior causes) but also thought about the new implications of statistics in legal frameworks (Bertrand 1889), as well as producing philosophical thoughts (p. 23):

> La Physique, l'Astronomie, les phénomènes sociaux, semblent, dans plus d'un cas, régis par le hazard. Peut-on comparer la pluie ou le beau temps, l'apparition ou l'absence des étolies filantes, la santé ou la maladie, la vie ou la mort, le crime ou l'innocence à des boules blanches ou noires tirées d'una meme urne? Le meme désordre apparait dans les details, cache-t-il la meme uniformité dans les moyennes? Retrouvera-t-on dans les écarts les traits connus et la physionomie des effets du hazard? (p. 31, from now) Les lois du hazard son invariables, ce sont les conditions du jeu qui changent. Poisson, pour les plier à tous les accidents, a cru completer l'oeuvre de Bernoulli en énonçant sa *loi des grands nombres*.

Bertrand accepted the overall presence of hazard in human domains but at the same time offered a conceptual tool, statistics, to deal with it. At a certain point, he considered that the Medieval debate about the universals had reached a new level when Quetelet was able to define the work "man" (*homme*) apart from particular men, considered accidents, (Bertrand 1889: 41): the *home moyen* (the average man). This *home moyen* has an *àme moyenne* (average soul, ibid, p. 43) and suffers from a "maladie moyenne que la Statistique révelerait pour lui" (an average illness that will be discovered by Statistics). As it is usual, thinkers faced with the deep

---

[15]Very surprisingly, the military swiftly accepted Bayesian techniques during the Second World War and Cold War, as well as related agencies such as NASA. And this is valid for American and European efforts: In 1979, NATO organization held a symposium to encourage Bayesian application for real conflicts.

implications about statistical analysis of Nature need to justify conceptually new visions about the world and the ways by which we know it. Without computational power, Bayesian methods were difficult and hard to apply, and all this changed slightly from the beginnings of the 1960s and completely since the 1980s. The other impediment, how to apply scientifically the subjective interpretation of probability, despite it being invented in the 1930s by Ramsey and De Finetti, did not become well-known and applied until the 1950s with the work of Savage and Lindley.[16] This new interpretation solved the main problems of the older "degree of rational belief" interpretation and gave Bayesian ideas a new space into academic research.[17]

References

Bayes, T. (1763). An essay towards solving a problem in the doctrine of chances. *Philosophical Transactions of the Royal Society of London, 53*, 370–418.

Bellhouse, D. R. (2002). On some recently discovered manuscripts of Thomas Bayes. *Historia Mathematica, 29*, 383–394.

Bellhouse, D. R. (2004). The reverend Thomas Bayes, FRS: A biography to celebrate the tercentenary of his birth. *Statistical Science, 19*(1), 3–43.

Bertrand, J. L. F. (1889). *Calcul des probabilités*. Paris: Gauthier-Villars et fils.

Bernardo, J. M. (2011). Modern Bayesian inference foundations and objective methods. In D. M. Gabbay, P. Thagard & J. Woods (Eds.), *Handbook of the Philosophy* (Vol. 7, pp. 263–306). Oxford (UK): Philosophy of Statistics, Elsevier.

Bertsch McGrayne, S. (2011). *The theory that would not die how Bayes' rule cracked the enigma code hunted down Russian submarines and emerged triumphant from two centuries of controversy*. USA: Yale University Press.

Dale, A. I. (1999). *A history of inverse probability from Thomas Bayes to Karl Pearson*. New York: Springer-Verlag.

Earman, John. (1992). *Bayes or bust?*. Cambridge, MA: MIT Press.

Efron, B. (2012). A 250-year argument: Belief, behavior, and the bootstrap. *Bulletin American Mathametical Society, 50*, 129–146.

Fayé, H. (1884). *Sur l'origine du monde, théories cosmogoniques des anciens et des modernes*. Paris: Gauthier-Villars et fils.

Fischer, H. (2010). *A history of the central limit theorem: From classical to modern probability theory (sources and studies in the history of mathematics and physical sciences)*. Germany: Springer.

Galavotti, M. C. (2003). Harold Jeffreys' probabilistic epistemology: Between logicism and subjectivism. *British Journal for the Philosophy of Science, 54*, 43–57.

Gillies, D. (1987). Was Bayes a Bayesian? *Historia Mathematica, 14*, 325–346.

Gillies, D. (2000). *Philosophical theories of probability*. UK: Routledge.

---

[16]Curiously, Lindley gave the name to a paradox, called Lindley's paradox, that put Bayesian and fequentists into the same bottleneck regarding a hypothesis testing problem that gives different results for certain choices of the prior distribution.

[17]I thank to Prof. D. Gillies his generous and detailed conceptual suggestions about this section of the book.

Hald, A. (2007). Laplace's theory of inverse probability, 1774–1786. In *A History of Parametric Statistical Inference from Bernoulli to Fisher, 1713–1935,* Part II (pp. 33–46).

Laplace, P.-S. (1814). *Essai philosophique sur les probabilités.* Paris: Courcier. References to this work have been made from *Ouvres complètes de Laplace,* (Vol. 7). Paris: Gauthier-Villars, 1886.

Laplace, P.-S. (1774) Mémoire sur la probabilité des causes par les *événements.* In *Mémoires de mathématique et de physique presentés à l'Acedémie royale des sciences, par divers savants, & lûs dans ses assemblées* (Vol. 6. pp. 621–656*).* Reprinted in Laplace's Œuvres complètes, t. VI: 27–65). Downloadable at: http://gallica.bnf.fr/ark:/12148/bpt6k77596b/f32

Reichenbach, H. (1952). Current epistemological problems and the use of a three-valued logic in quantum mechanics. In M. Reichenbach & S. Cohen (Eds.), *Dedicated to Erich Regener on his seventieth birthday, Hans Reichenbach. Selected writings, 1909–1952* (Vol. 2). London: D. Reidel Publishing Company.

Sewell, G. (2002). *Quantum mechanics and its emergent macrophysics.* USA: Princeton University Press.

Stigler, S. M. (1986). Laplace's 1774 memoir on inverse probability. *Statistical Science, 1*(3), 359–363.

Vallverdú, J. (2011a). Bayesian versus Frequentist statistical Reasoning. In Lovric, M. (Ed.), *International Encyclopedia of Statistical Science* (pp. 133–135).

Chapter 4
A Conceptual Reply to Reverend Bayes: The Frequentist Approach

Abstract The response to subjective probabilities of the Bayesian approach was frequentism, that is, the analysis of long-run series of frequencies of an event from which came the possibilities to extract statistical data. Frequentism became the dominant view in scientific practices during most of the twentieth century. This academic view was espoused by several authors, like Pearson, Fisher, Gosset, and Neyman–Pearson, not all of them agreeing about the best ways to perform this statistical approach. The main ideas and internal debates are analyzed here.

Keywords Frequentism · Frequency · Long-run series · *p*-value · Pearson · Fisher · Neyman–Pearson · Null hypotheses

In all good stories, there is a strong character fascinating and charming in equal parts because of its association with the "dark side." This blend of a true, innocent, and brave hero confronted with a bad and not-smart-enough guy gives an irresistible smell of all the histories of the past. In our story, frequentists occupy this dark area, perhaps not for conceptual reasons, because frequentism is really useful and has provided successful tools in several fields and/or domains, but by the intransigent and aggressive attitude toward a different methodology, Bayesianism, that has demonstrated basics for very important scientific advances of the twenty-first century. Let me explain how and why frequentists acquired power inside the statistics communities and the works of their leading experts. At the same time, we will see how Bayesianism evolved.

First of all…what does "frequentism" mean? According to the definition provided by the Oxford English Dictionary:

> frequentist. [f. *frequent*-, stem of adjs., etc., related to FREQUENCY + -IST.]
> One who believes that the probability of an event should be defined as the limit of its relative frequency in a large number of trials. Also *attrib.* or as *adj.*

Therefore, we are talking about people who search relative frequencies in a large number of trials, that is, from facts to mathematical quantification without being exhaustive of all the possible cases (otherwise statistics would not be necessary). With a typical example, it will be easy to understand: coin tossing. A coin has two sides, and when you throw it into the air, once it lands the coin can be found in …

three positions: head, tail, and edge. It is true that most times, although you cannot discard the coin landing on its edge so happily because it can really happen, the coin will land head or tail; so, we are considering two possible outcomes. If the coin is perfectly balanced (something not possible unless both sides had exactly the same design), there are no biases into the throwing process and the surrounding variables (wind, temperature, floor shape…) are not different each time; the coin will land only as head or tail. So, if we consider a 100 % of a range of possibilities, 50 % of the time it will land on heads and 50 % on tail. Ancient Romans used this process to let the Gods decide between things and called it *capita aut navia*, the two usual signs present in coins.[1] And in the twenty-first century, it is still a practice used at football stadiums to decide the initial side of the field for each team. Isaac Asimov wrote a wonderful short story in 1961 called "The Machine that Won the War" following this idea. But…are we sure that each coin tossing will offer us a result heads and tails with equiprobability? This is precisely what a French naturalist and mathematician asked himself in the eighteenth century. He was Georges-Louis Leclerc, Comte de Buffon, and a very thorough guy who was able to toss a coin 4040 times trying to obtain greater evidence toward the equiprobability of this action. Heads came up on 2048 tosses (50.693 %), so it was very close. Well, to be honest, Buffon himself did not do it, he asked a kid to do it, as he explained in his *Essai d'arithmétique morale* (1777). If Buffon pursued exactitude, around 1900, Karl Pearson tossed a coin 24,000 times, which can be described close to an obsessive behavior, or a true statistician's mood. He obtained 12,012 heads (50.05 %). Finally, and only explained by the inexistence of leisure activities for prisoners during the WWII, the South African mathematician John Kerrich tossed a coin 10,000 times with heads coming up 5067 times (50.67 %). While visiting Copenhagen, Kerrich was caught up during the Denmark's Nazi invasion and imprisoned in the Frøslev Prison Camp (in Jutland area and a few km north of the German border). After the war, in 1946, Kerrich published the book *An Experimental Introduction to the Theory of Probability*. From the data provided by Kerrich in 1946 and their graphic made by Freedman et al. (1979), we find the distribution of toss results. It tends to be equiprobable on a long-run procedure. In a nutshell, this is the idea of frequentism: to obtain *a large number of trials* from which emerged the relative frequency of an event. Now, you can understand the main difference with bayesianism: no presence in frequentism of personal prior beliefs in this process.[2]

---

[1]From Smith et al. (1890): "CAPITA AUT NAVIA head or tail, the name of a game at 'pitch and toss,' derived from the fact that early *as* had on one side a double-faced Janus, on the other the prow of a ship. See cut of *as* on p. 202. (*Macr. 1.7, 22*; Fest. s. v. Navia, p. 169 M.)".

[2]Well, we could discuss how personal beliefs cannot affect a frequentism experiment in the selection of the good variables that can affect or not affect the experiment or, even, the meaning of "a large number of trials," something completely subjective and that can vary from discipline to discipline or even from laboratory to laboratory. For a nice and opposed vision of my ideas on this topic, read Gillies (200): 152–153. Anyhow, the selection of the basic range of trials is absolutely subjective.

Between 1885 and 1935, the Statistical Enlightenment happened (according to Stigler 2012), which changed the nature and evolution of statistics as a discipline as well at its conceptual core. It is time now to know some of the stronger defenders of this statistical school. In 1900, an English mathematician, Karl Pearson, wrote a paper that was the first step toward modern mathematical statistics. Pearson was a talented and curious researcher, Germanist, Literature expert, Philosopher[3], and Mathematician, who was also involved in the quantitative new approaches to biology made by evolutionary experts at the end of the nineteenth century, working closely with eugenicist Francis Galton (for whom he was also the official biographer) with whom (and Weldon) he created and edited the first statistics journal *Biometrika*, for 35 years. The empirical (statistical) foundations of modern biology favored the adoption of frequentist approaches, but after some decades, later this dominance was altered by new tools and ideas. Pearson was a leading figure who established a real school of statistics, being at the same time the main inspiration for the discipline in the twentieth century.[4] It was precisely Galton's interests in regression and correlation in psychology, heredity, and anthropology that led Pearson to the intense study of them during the period of 1891–1900 (Plackett 1983). As a consequence of it, in 1900, his paper appeared on the chi-square test of goodness of fit: "On the criterion that a given system of deviations from the probable in case of a correlated system of variables is such that it can be reasonably supposed to have arisen from random sampling." In this research, he offered a system of chance distribution, called the *chi-square test* (χ^2), and laid the foundation stone of modern frequentist statistics. It is a hypothesis test using normal approximation for discrete data, and he developed a system of chance distributions, each obtainable via a variation in the parameters appearing in a "generalized probability curve," a formula that generalizes the normal distribution. A few years later, in 1911, he founded the world's first university statistics department at University College London. Under the figure of Karl Pearson, we find the figure of

---

[3]Of great philosophical interest is his *The Grammar of Science* (1892), a book about which the young Albert Einstein was enthusiastic, and also received critics from Lenin on the debate between materialism and idealism, Lenin, V.I. (1909) *Materialism and Empirio-Criticism. Critical Comments on a Reactionary Philosophy*, Ch. 5, Sect. 2. Lenin described him as a "machian." Read this section at: http://www.marxists.org/archive/lenin/works/1908/mec/five2.htm, accessed on August, 13, 2013. At the same time, we need to explain that Pearson wrote on epistemological issues: For example, he considered that knowledge came from sensations and that probability tried to find invariability among these groups pf sensations (shared with other individuals as a "sameness experience"). Lenin quoted him on this topic at V.I. Lenin (1908) *Materialism and Empirio-Criticism*, Critical Comments on a Reactionary Philosophy, Chapter One: The Theory of Knowledge of Empirio-Criticism and of Dialectical Materialism. (1) Sensations And Complexes Of Sensations. There is a strong connection between his notions of science and the future of human societies (Norton 1978).

[4]Pearson is also the longest "sleeping beauty" in the history of science; he wrote "On lines and planes of closest fit to systems of points in space" in 1901, but his ideas gained approval only in 2002. It indicates that changes in society and advances in understanding can breathe new life into sometimes long-forgotten science papers. See the last research on this topic at Ke et al. (2015).

a chemist and mathematician who worked for brewery industries: William Sealy Gosset. Gosset was an employee of the Guinness firm and employed his knowledge to the statistical selection of the best yielding varieties of barley and other chemical questions related to brewery techniques. He spent two terms (1906–1907) in the biometrical laboratory of Pearson where they became friends as Gosset decided the importance of small samples' statistical treatment. Due to industrial restrictions of Guinness industrial secrets, Gosset was not allowed to publish his results directly, even after demonstrating to Guinness heads that his research was philosophical and mathematical with no danger to Guinness drinks' secret industrial processes, and therefore, he was forced to use a pseudonym, *Student*, with which he wrote several papers, most of them published in *Biometrika*. His greatest contribution, based on the analysis of small samples, was the so-called the Student's t-distribution (Student 1908), and he referred in this paper to the distribution as the "frequency distribution of standard deviations of samples drawn from a normal population." Fisher was in that time not truly interested in small samples because his analysis was focused on the contrary: big data samples (at least for an early twentieth-century scientist).

R.A. Fisher is the next historical actor we will consider in this chapter. Again, we find a personality with strong interests in mathematics as well as in biology, especially evolutionary theory[5] and eugenics (Fisher was one of the founders of the Eugenics Society of the University of Cambridge, together with John Maynard Keynes, according to Howie 2002: 52).[6] Fisher was a talented, hardworker, and devoted researcher who considered Bayesianism as the main mistake and error of statistics and embraced philosophical Laplacian of a full deterministic world described by statistics, which was the result of objective properties of the world, not from subjective performing of the mind. Having been granted a studentship in physics (1912–1913), Fisher dedicated his graduate research in Cambridge to quantum physics (with Physicist James Jean) and error theory (with F.J.M. Stratton, who had applied it with T.B. Wood to agriculture, see Grattan-Guinness 1994). As a consequence of this background, he understood the meaning of Heisenberg's principle of uncertainty, although from his objectivist approach to stochastic events. As he explicated in his paper of 1922, "On the Mathematical Foundations of Theoretical Statistics," "we may agree wholly with CHRYSTAL at inverse probability is a mistake (perhaps the only mistake to which the mathematical world has so deeply committed itself), there yet remains the feeling that such a mistake would

---

[5]According to Loucã (2008: 3), Fisher's model of Mendelian populations was metaphorized from the molecular models of statistical mechanics applied to gases.

[6]R.C. Punnet, and Leonard and Horace Darwin, both sons of Charles Darwin, were also members of this society. The friendship between Leonard Darwin and R.A. Fisher was very intense, sharing their interest in eugenics and evolutionary ideas. Besides, Leonard gave support to Fisher at the beginning of his career. Also Pearson was close to these circles, in this case friend of Ida Darwin, wife of Horace Darwin, as some correspondence shows (http://www.eugenicsarchive.org/html/eugenics/static/images/2140.html, accessed on August 12th, 2014). Again, eugenics was the common point.

not have captivated the minds of LAPLACE and POISSON if there had been nothing in it but error" (p. 311), and for him, probability was something very clear: "When we speak of the probability of a certain object fulfilling a certain condition, we imagine all such objects to be divided into two classes, according as they do or do not fulfil the condition This is the only characteristic in them of which we take cognisance. For this reason probability is the most elementary of statistical concepts. It is a parameter which specifies a simple dichotomy in an infinite hypothetical population, and it represents neither more nor less than the frequency ratio which we imagine such a population to exhibit" (p. 312). Inverse probability was part of an "obscure" era of statistics that needed to be eradicated. He wrote two books that offered his ideas about how statistics should be performed as well as how experiments should be designed: (1925) *Statistical Methods for Research Workers* and (1935) *The Design of Experiments*. Both works were extremely successful and were adopted by several disciplines and reprinted from time to time. Fisher was the real creator of the notion of a statistical model, refining the concepts of "variables" and "parameters" as well as establishing the distinction between sample and population. In the field of statistics, Fisher was terribly prolific and his influence was deep as well as full of debates with other experts in the field. With Neyman and Pearson, he maintained private as well as public debates (see Inman 1994), but he can be remembered by crucial contributions such as the ANOVA,[7] the method of maximum likelihood, the fiducial inference, or the derivation of various sampling distributions. Fiducial inference has a philosophical flavor that is impossible to avoid any expert in statistics. "Fiducial" comes from the Latin word *fiducia* for "faith," and fiducial inference can be interpreted as an attempt to perform inverse probability without calling on prior probability distributions, an idea that received several counterexamples very soon and that have not reached general acceptance.[8] Additionally, Fisher published one of the world's first scientific papers, in fact, the first one in the field of biology using computer calculations, which were done on EDSAC (Electronic Delay Storage Automatic Calculator), the first fully operational and practical stored-program computer. Fisher suggested a genuine problem, the solution of a second-order nonlinear differential equation with two-point boundary condition relating gene frequencies, which was solved by D.J. Wheeler in April 1950 and was published with due acknowledgements later in the year in Biometrics (Fisher 1950), making some progress toward establishing the credibility of electronic computing.

---

[7]Later, in 1952, Kruskal and Wallis published a paper in which they provided a nonparametric equivalent of the ANOVA. And John Tukey designed an ANOVA in 1951 a posteriori multiple comparison tool: the Tukey's HSD (for honestly significant difference test). Tukey's multiple comparison test is one of several tests that can be used to determine which means among a set of means differ from the rest. There are more multiple comparison tests, including Scheffe's test and Dunnett's test.

[8]Curiously, in the early 1930s, most statisticians regarded fiducial probability and Neyman's confidence intervals as synonymous (Louçã 2008: 22).

The third historically important figure in frequentism is Jerzy Neyman, a Polish mathematician who was the first to introduce the modern (and *beautiful*, according to Cousins 1995) concept of a confidence interval, one of the methods of confidence estimation. His ideas made the birth of a new theoretical statistics paradigm possible deriving optimal statistical procedures as solutions to clearly stated mathematical problems. His application of these ideas to hypothesis testing, estimation by confidence intervals, and survey sampling changed the history of statistics. Despite being strongly influenced by Karl Pearson's *Grammar of Science* (Lehmann 1994: 397), until 1926, he favored Bayesian approaches from the belief that any theory would have to involve statements about the probabilities of various alternative hypotheses and hence an assumption of prior probabilities. But after his joint work with Karl Pearson and the influence of von Mises's book *Wahrscheinlichkeit, Statistik und Wahrheit* (1928), Neyman turned himself into a radical frequentist. His necessity of "purity" led him to abandon the notion of inductive reasoning, pregnant to his eyes of subjective data (or "beliefs"), and to talk about "comportement inductif" or inductive behavior, that is, that statistics is to be used not to extract "beliefs" from experience, but as a guide to appropriate action. In a paper presented at the International Congress on the Philosophy of Science in 1949, Neyman gave his ideas:

> Why abandon the phrase 'inductive reasoning' in favor of 'inductive behavior'?" As explained in 1937, the term inductive reasoning does not seem appropriate to describe the new method of estimation because all the reasoning behind this method is clearly deductive.[9] Starting with whatever is known about the distribution of the observable variables X, we deduce the general form of the functions f(X) and g(X) which have the properties of confidence limits. Once a class of such pairs of functions is found, we formulate some properties of these functions which may be considered desirable and deduce either the existence or non-existence of an "optimum" pair, etc. Once the various possibilities are investigated we may decide to use a particular pair of confidence limits for purposes of statistical estimation. This decision, however, is not 'reasoning'. This is an act of will just as the decision to buy insurance is an act of will. Thus, the mental processes behind the new method of estimation consist of deductive reasoning and of an act of will. In these circumstances the term 'inductive reasoning' is out of place and, if one wants to keep the adjective 'inductive', it seems most appropriate to attach to it the noun 'behavior'.

The strong relationship between his philosophical ideas on the nature of scientific methodology and the role of statistics is completely connected in Neyman's mind. Even working with Karl Pearson, he was not satisfied with his laboratory, which he considered old-fashioned, neither by his lack of modern mathematical knowledge (which led to a misunderstanding between them, because Pearson was not able to understand the differences between the ideas of independence and lack of correlation, and this led to Neyman's decision to move to a new research

---

[9]According to Gillies (personal communication), it can be affirmed, and it is worth noting that Popper was also criticizing the notion of inductive reasoning in the 1930s. Generally, classical (or frequentist) statistics was very much in agreement with Popper's anti-inductivist methodology of conjectures and refutations, except for the fiduciary argument which is definitely inductivist in character.

place[10]). Regardless, this confrontation was not a problem for the posterior cooperative debates with Egon Pearson, Karl Pearson's son, who occupied (half of)[11] the vacant chair left by his father at the university. Egon and Neyman worked together trying to solve a big problem that had emerged from Fisher and Student studies: The small sample tests showed an ad hoc nature, and this resulted in offense for any frequentist expert. After some ideas exchanges with Student, Egon and Neyman created what has been called the "Neyman–Pearson theory of hypothesis testing." In 1928, they published "On the Use and Interpretation of Certain Test Criteria for Purposes of Statistical Inference: Part I." Let them convey their ideas:

> One of the most common as well as most important problems which arise in the interpretation of statistical results, is that of deciding whether or not a particular sample may be judged as likely to have been randomly drawn from a certain population, whose form may be either completely or only partially specified (...) The sum total of the reasons which will weigh with the investigator in accepting or rejecting the hypothesis can very rarely be expressed in numerical terms. All that is possible for him is to balance the results of a mathematical summary, formed upon certain assumptions, against other less precise impressions based upon a priori or a' posteriori considerations. The tests them- selves give no final verdict, but as tools help the worker who is using them to form his final decision; one man may prefer to use one method, a second another, and yet in the long run there may be little to choose between the value of their conclusions. What is of chief importance in order that a sound judgment may be formed is that the method adopted, its scope and its limitations, should be clearly understood, and it is because we believe this often not to be the case that it has seemed worth while to us to discuss the principles involved in some detail and to illustrate their application to certain important sampling tests (...) [and at the conclusions section, italics are mine] *The system adopted will provide a numerical measure*, and this must be coordinated in the mind of the statistician with a clear understanding of the process of reasoning on which the test is based. We have endeavored to connect in a logical sequence several of the most simple tests, and in so doing have found it essential to make use of what R. A. Fisher has termed "the principle of likelihood." The process of reasoning, however, is necessarily an individual matter, and we do not claim that the method which has been most helpful to ourselves will be of greatest assistance to others. It would seem to be a case where each individual must reason out for himself his own philosophy.

This paper contained the seeds of what was later improved and called the Neyman–Pearson theory of hypothesis testing (NPTHT). Egon Pearson was son of Karl Pearson and succeeded him as professor of statistics at University College London as well as editor of the journal *Biometrika*. With Neyman, Egon created the NPTHT.[12] With this approach, they opposed Fisher face-on; it was Fisher who

---

[10]When Neyman moved to the University of California at Berkeley, he transformed that place into an anti-Bayesian powerhouse.

[11]The second half of K. Fisher divided position was given to R.A. Fisher, a next author to be covered in this chapter. Despite the objections of Karl Pearson, his laboratory was divided into separate departments of statistics and eugenics. His son became head of the new Department of Statistics, while R.A. Fisher was elected as Galton Professor of National Eugenies (Inman 1994: 4).

[12]Curiously, George Box, a former student of Pearson, became Bayesian and even married one of Fisher's daughters...some years after he divorced her, because his wife had inherited a temper much like her father.

defended that only the null hypothesis needed to be tested in a binomial procedure. On the contrary, Neyman and Pearson defended that after defining a hypothesis, one could consecutively test multiple alternatives against this hypothesis and that there were different kinds of errors that could be detected (Type I and Type II, respectively, the incorrect rejection of a true null hypothesis and the rejection of a false null hypothesis, that is, false positives and false negatives).[13]

We will analyze this debate between Fisher and Neyman–Pearson, taking null hypothesis, created by Fisher, as a central concept. When somebody tries to establish a causal relationship between two measured phenomena, there can be difficulties to ascertain whether the relationship is spurious or has a direct relationship between some cause and some effect. A false-positive result could show us a relationship that does not exist.[14] With the null hypothesis, this can be partially avoided, and the idea is simple: You define a hypothesis that defends that there is no relationship between two phenomena (or sets of data); if it is true, then you prove a false relationship and can guide your efforts toward a different direction. Philosophically, it is close to Popper's notion of falsifiability (Popper 1934), despite the true statistical nature of Fisher's approach, and again has a deep relationship with the problem of induction.[15] Fisher conceived this concept as a practical tool for agricultural research studies. In 1926, he affirmed:

> In the investigation of living beings by biological methods, statistical tests of significance are essential. Their function is to prevent us being deceived by accidental occurrences, due not to causes we wish to study, or are trying to detect, but to a combination of many other circumstances which we cannot control. An observation is judged significant, if it would rarely have been produced, in the absence of a real cause of the kind we are seeking. It is common practice to judge a result significant, if it is of such a magnitude that it would have been produced by chance not more frequently than once in twenty trials. This is an arbitrary, but convenient, level of significance for the practical investigator, but it does not mean that he allows himself to be deceived once every twenty experiments. The test of significance only tells him what to ignore, namely all experiments in which significant results are not obtained. He should only claim that a phenomenon is experimentally demonstrable when he knows how to design an experiment so that it will rarely fail to give a significant result. Consequently, isolated significant results which he does not know how to reproduce are left in suspense pending further investigation. (p. 189)

For Fisher, a null hypothesis could be potentially rejected or disproved on the basis of related data, and this led to the *p-values* notion. But Neyman and Pearson

---

[13]There is also a Type III error: when you get the right answer to the wrong question. This is sometimes called a Type 0 error. This error arises from a two-sided test, when one side is erroneously favoured although the true effect actually resides on the other side. It is not a false positive but a crossed causal relationship. See Schwartz and Carpenter (1999); a more recent analysis in Heinz and Waldhoer (2012).

[14]This obvious fact should introduce modesty in statisticians as well as a careful epistemological attitude, (Boffetta et al. 2008; Blair et al. 2009).

[15]As Meehl (1990), p. 110, explained quoting the words of American philosopher Morris Raphael Cohen: "All logic tests are divided into two parts. In the first part, on deductive logic, the fallacies are explained; in the second part, on inductive logic, they are committed." Induction is one of the oldest and more conflictive problems in the history of philosophy.

made a different approach to this idea[16]: They suggested using two complimentary hypotheses (called null and alternative).[17] For a deep analysis of this debate on null hypothesis significance testing, see Robinson and Wainer (2001) as well as Louçãa (2008): 10. Curiously, very recently, the journal *Basic and Applied Social Psychology* published an editorial banning officially null hypothesis significance testing procedures (NHSTP): They wanted to remove all vestiges of the NHSTP (p-values, t-values, F-values, statements about "significant" differences) and decided that Bayesian methods were more interesting (despite the problems generated by the Laplacian assumption of equiprobability).[18] Previously, researchers like Wagenmakers (2007) tried to avoid p-values because of the imaginary statements[19] (implies the faith about some value obtained *if* the study was repeated an *infinite number of times*) and suggested a "Bayesian model" that does not require the specification of that he called Bayesian information criterion (BIC).

In 1930, Neyman wrote to Pearson telling him that he has discovered a rigorous argument in favor of the likelihood method, what led him to create the "Fundamental Lemma" of Neyman–Pearson theory on NPTHT. Paradoxically, Imre Lakatos, a disciple of Popper, considered Neyman–Pearson test of hypothesis as "resting completely on methodological falsificationalism" (Lakatos 1978: 25), although both developed this idea before and independently to Popper.

Anyhow, Fisher's test of significance has arrived upon our days mixed with Neyman–Pearson hypothesis test, despite the several debates about their similitude or difference (Lehmann 1993; Lenhard 2006). Abraham Wald was an Austro-Hungarian mathematician who wrote a paper on statistical theory in 1939 and later created the maximin model. Curiously, during WWII research on statistical ways to improve fighter airplanes, armor was conducted leading to the definition of the survivorship bias. He died unexpectedly in an airplane crash in India during a conference tour. Wald's theory was heavily criticized by Fisher, but Neyman decided to defend his legacy. A young frequentist, Howard Raiffa, was hired by Columbia University to teach Wald's course and he studied Wald's book intensively each night (keeping only a day ahead of his students). Raiffa soon realized the power and depth of Wald's methods and ideas, especially for decision-making situations, going beyond the classic and static data analysis.

---

[16]See the acid analysis of Nuzzo (2014). As she points, p. 151: "Neyman called some of Fisher's work mathematically "worse than useless"; Fisher called Neyman's approach "childish" and "horrifying (for) intellectual freedom in the west." As Bertsch (2011: 46) compiles from his contemporaries, Fisher was aggressive, unpolite, fiery tempered. About p-values, Nuzzo is even sardonic (p. 150): "P values have always had critics. In their almost nine decades of existence, they have been lik-ened to mosquitoes (annoying and impossible to swat away), the emperor's new clothes (fraught with obvious problems that everyone ignores) and the tool of a "sterile intellectual rake" who ravishes science but leaves it with no progeny. One researcher suggested rechristening the methodology "statistical hypothesis inference testing", presumably for the acronym it would yield...and she meant SHIT....

[17]About this debate, read the precise paper of Gillies (1971).

[18]Trafimov and Marks (2015).

[19]Head talks about "p-hacking." Read Head et al. (2015).

Together with Schlaifer, from Havard University, Raiffa worked intensively on Bayesian decision in economic contexts. In 1954, Leonard Jimmie Savage wrote his influential book *The Foundations of Statistics*, the first modern book on Bayesian statistics[20], and in 1959, Schlaifer wrote the book *Probability and Statistics for Business Decisions*, the first textbook written entirely from the Bayesian point of view. Trying to improve the implementation and reception of Bayesian tools, Raiffa[21] and Schlaifer introduced decision trees, tree-flipping, and conjugate priors. All of them contributed to introduce Bayesian ideas into several disciplines related to decision-making procedures (especially in decision making in uncertainty contexts). They both contributed to the Bayesian revival of the 1960s.

Until here, we have seen the main figures of frequentist approaches (without considering Fisher's likelihoodism as a completely different school), and in the next chapter, we will analyze the debate and evolution of both approaches during the second half of the twentieth century.

References

Blair, A., et al. (2009). Epidemiology, public health and the rhetoric of false positives. *Environmental Health Perspectives, 117*(12), 1809–1813.

Boffetta, P., et al. (2008). False positive results in cancer epidemiology: A plea for epistemological modesty. *Journal of the National Cancer Institute, 100*, 988–995.

Cousins, R. D. (1995). Why isn't every physicist a Bayesian? *American Journal of Physics, 63*(5), 198–410.

Fienberg, S. E. (2008). The early statistical years: 1947–1967 a conversation with Howard Raiffa. *Statistical Science, 23*(1), 136–149.

Fisher, R. A. (1922). On the mathematical foundations of theoretical statistics. *Philosophical Transactions of the Royal Society of London. Series A, 222*, 309–368.

Fisher, R. A. (1950). Gene frequencies in a cline determined by selection and diffusion. *Biometrics, 6*(4), 353–361.

Freedman, D. A., Pisani, R., & Purves, R. A. (1979). *Statistics.* NY: W.W. Norton.

Gillies, D. A. (1971). A falsifying rule for probability statements. *British Journal for the Philosophy of Science, 22*, 231–261.

Grattan-Guinness, I. (Ed.). (1994). *Companion encyclopaedia of the history and philosophy of the mathematical sciences.* London: Routledge.

---

[20]Savage was a brilliant and ironic researcher, and at the same time, he showed deep interests in philosophical aspects of statistical concepts. For example, in "The Foundations of statistics Revisisted" he wrote: "Fisher's school, with its emphasis on fiducial probability-a bold attempt to make the Bayesian omelet without breaking the Bayesian eggs—may be regarded as an exception to the rule that frequentists leave great latitude for subjective choice in statistical analysis" (1961, p. 578), published at Berkeley Symposium on Mathematical Statistics and Probability. He considered that "Personal probability at present provides an excellent base of operations from which to criticize and advance statistical theory." Bayesianism had with him a strong leader and deep thinker.

[21]Curiously, Raiffa wrote most of his ideas as books, with some scarce and minor papers. And, again, his beginnings were related to military war efforts. Read the wonderful paper about his life (as an interview) made by Fienberg (2008).

Head, M. L., Holman, L., Lanfear, R., Kahn, A. T., & Jennions, M. D. (2015). The extent and consequences of P-Hacking in science. *PLoS Biology, 13*(3), e1002106.

Heinz, H., & Waldhoer, T. (2012). Relevance of the type III error in epidemiological maps. *International Journal of Health Geographics, 11*(34), 1–9.

Howie, D. (2002). *Interpreting probability: Controversies and developments in the early twentieth century.* Oxford: OUP.

Inman, H. F. (1994). Karl Pearson and R.A. Fisher on statistical tests: A 1935 exchange from nature. *The American Statistician, 48*(1), 2–11.

Ke, Q., Ferrara E., Radicchi, F., & Flammini, A. (2015). Proceedings of the National Academy of Sciences USA. http://dx.doi.org/10.1073/pnas.1424329112. Accessed in May 26, 2015.

Kruskal, W. (1952). Use of ranks in one-criterion variance analysis. *Journal of the American Statistical Asociation, 47*(260), 583–621.

Lakatos, I. (1978). *The methodology of scientific research programmes: Philosophical papers* (Vol. 1). Cambridge: Cambridge University Press.

Lehmann, E. L. (1993). The Fisher, Neyman-Pearson theories of testing hypotheses: One theory or two? *Journal of the American Statistical Association, 88*(424), 1242–1249.

Lehmann, E. L. (1994). *Jerzy Neyman 1894–1981. A biographical memoir.* Washington D.C.: National Academy of Sciences.

Lenhard, J. (2006). Models and statistical inference: The controversy between Fisher and Neyman–Pearson. *British Journal for the Philosophy of Science, 57*, 69–91.

Louçã, F. (2008). The widest cleft in statistics—How and why Fisher opposed Neyman and Pearson. *Working Papers WP 02/2008/DE/UECE.* Lisbon: ISEG.

McGrayne, S. B. (2011). *The theory that would not die. How Bayes' rule cracked the enigma code, hunted down Russian submarines, and emerged triumphant from two centuries of controversy.* USA: Yale University Press.

Meehl, P. E. (1990). Appraising and amending theories: The strategy of Lakatosian defense and two principles that warrant using it. *Psychological Inquiry, 1*, 108–141.

Neyman, J. (1928). On the use and interpretation of certain test criteria for purposes of statistical inference: Part I. *Biometrika, 20A*(1/2), 175–240.

Neyman, J. (1949). Foundations of the general theory of estimation. *Actualites Scientifiques et Industreilles, 1951*(1146), 85.

Norton, B. J. (1978). Karl Pearson and statistics: The social origins of scientific innovation. *Social Studies of Science, 8*(1), 3–34.

Nuzzo, R. (2014). Statistical errors. *Nature, 506*, 150–152.

Pearson, K. (1900). On the criterion that a given system of deviations from the probable in case of a correlated system of variables is such that it can be reasonably supposed to have arisen from random sampling. *Philosophical Magazine Series, 50*(302), 157–175.

Plackett, R. L. (1983). Karl Pearson and the chi-squared test. *International Statistical Review, 51* (1), 59–72.

Popper, K. (1934). *Logik der Forschung: Zur erkenntnistheorie der modernen naturwissenschaft.* Vienna: Springer.

Robinson, D. H., & Wainer, H. (2001). *On the past and future of null hypothesis significance testing. Research report RR-01-24.* Princeton: ETS.

Savage, L. J. (1961). Berkeley symposium on mathematical statistics and probability. In *Proceedings of the Fourth Berkeley Symposium on Mathematical Statistics and Probability* (Vol. 1, pp. 575–586). Berkeley, California: University of California Press.

Schwartz, S., & Carpenter, K. M. (1999). The right answer for the wrong question: Consequences of type III error for public health research. *American Journal of Public Health, 89*(8), 1175–1180.

Smith, W., William Wayte, L. L. D., & Marindin, G. E. (1890). *A dictionary of greek and roman antiquities.* London: John Murray.

Stigler, S. M. (2012). Karl Pearson and the rule of three. *Biometrika, 99*(1), 1–14.
Student. (1908). The probable error of a mean. *Biometrika, 6,* 1–25.
Trafimov, D., & Marks, M. (2015). Editorial. *Basic and Applied Psychology, 37,* 1–2.
Wagenmakers, E. J. (2007). A practical solution to the pervasive problems of *p*-values. *Psychonomic Bulletin & Review, 14*(5), 779–804.
Wald, A. (1939). Contributions to the Theory of Statistical Estimation and Testing Hypotheses. *Annals of Mathematical Statistics, 10*(4), 299–326.

Chapter 5
The Coevolution, Battles, and Fights of Both Paradigms

Abstract During the second half of the twentieth century, the statistical arena experienced a turbulence of new debates: first, among frequentists, especially the classic such as K. Pearson, Neyman, E. Pearson, and the so-called likelihoodist R. Fisher, second among objective and subjective Bayesian, and finally, between Bayesian and frequentists. After exploring a demarcation process of all the involved schools, the strong divergences as well as the similarities between both main paradigms are posited.

Keywords Frequentism · Likelihoodism · Objective Bayesian · Subjective Bayesian · Experiment design · Mixture · Demarcation · Ad hoc rules · Regression · Mean · Controversy

Any debate among statistics experts about the nature and meaning of statistics rely on a previous philosophical view (even unconsciously). At the same time, they do not want to enter into the philosophical arena and try to justify themselves based on the evidence of their ideas or the power of their results, something quite childish. Bayesians are usually much more practical than frequentists in several research fields, considering their methods as problem-solving tools, while the latter feel a deep reluctance to admit "subjective values" into their protocols. Nevertheless, Bayesianism is not the realm of subjective-crazy priors[1] nor is frequentism the heaven of disinfected rationalism. In fact, frequentism is faced with several and

---

[1]Greenland (2006): 766 makes a strong and clear defense of the non-arbitrariness of subjective probabilities. At the same time, he claims against some specific views of Bayesian thinking that have become a major obstacle to the dissemination of Bayesian procedures. Subjectivity in prior distribution is minimized through basing prior information on defensible evidence and reasoning, and with accumulation of data differences of priors are solved by general consensus. The process of eliciting a prior distribution involves a dialogue between an expert and a statistically trained facilitator. Priors can be based on sound evidence and reasoned judgements, and they are not foolish claims from brainless guys!

serious accusations of subjectivism: *p-values*,[2] experiment design, ad hoc rules, arbitrary features of Neyman–Pearson tests, the difficulties caused by regression to the mean, and the relevance of stopping rules, among others. And another and more important question is, boundaries between these schools are not always so clear and this is in part a result of a practical use of statistics, beyond philosophical debates on the meaning of them.

5.1 Demarcation Problems

Until now, I have talked about frequentists and Bayesians, as clear delimitated groups, but this is not truly exact because there are more views and ideas that cannot strictly fit into these categories. The same authors implied into this classification would surely disagree about this point, but most of them can be defined as belonging to some of these different schools (Hawthorne 2011: 335):

I. Frequentists:

    (a) Classic: K. Pearson, Neyman, E. Pearson, Wald.
    (b) Likelihoodism: Fisher.

II. Bayesians[3]:

    (a) Objective Bayesians: Bayes (1763), Laplace (1814a, b), Keynes (1921), Jeffreys (1939), Carnap (1950), Edwin and Jaynes (1968), Fitelson (2001a, b), Williamson (2013).

---

[2]According to Greenland and Poole (2013), *p*-values and their corresponding frequentist (mis) interpretations can be better understood by correct Bayesian ideas. A mixture of both paradigms improves the practice of this technique. At the same time, we need to consider the limitations and conflictive nature of *p*-values, and we will exemplify it with fMRI studies...and salmon's brains. In 2008, a neuroscientist Craig Bennet was discussing with his coadviser George Wolford about false positives in fMRI multiple comparisons. Just to check their ideas, they decided to scan a dead salmon head and the look at the results. Incredibly, they found an empiric example about a real false positive and some non-sensical inferences that could be extracted from that scan. They presented the results as a poster at the *Human Brain Mapping Conference*, held in San Francisco. Submitted as an abstract, the poster was rejected, but when they wrote a paper with the results, the journal discussed intensively about their possible inclusion. The whole story can be found here: http://prefrontal.org/blog/2009/09/the-story-behind-the-atlantic-salmon/, accessed on July, 17, 2014. As the author remarked: "redefined significance thresholds with a specified cluster extent are a weak control to the problem of false positives in imaging data. Statisticians and methods researchers have argued about the need for multiple comparisons correction for some time. In just one figure the salmon data illustrates exactly why we need stronger controls for the false positive problem in fMRI."...obviously the were awarded with 2012 igNobel in neuroscience...The full paper here: http://pages.vassar.edu/abigailbaird/files/2014/06/bennett_salmon.pdf.

[3]There are a large number of types of Bayesians and speaking ironically, Good (1971) spoke of the existence of "46,656 kinds of Bayesians", depending on their attitude toward subjectivity in postulating priors. A little bit of humor can help to maintain a healthy debate.

(b) Subjective Bayesians: Ramsey (1931), de Finetti (1931), Savage (1954), Lewis (1980), Skyrms (1984), Howson and Urbach (1991), Bernardo[4] and Smith (1996), Howson (1997), Joyce (1992), Bradley (2012).[5]

Obviously, this is not an exhaustive list of all authors of possible classification under these categories, just a sketch for information purposes. It is important to remark that the rebirth of Bayesian models was the result of de Finetti's 1937 text, although to be honest, his writings, together with those of Ramsay and Jeffreys, lay unread until the Second Word War.[6] Despite the vigorous defense of Bayesianism by de Finetti, this view was not widely held and implemented by statisticians until the publication of Savage (1954), and under his influence, it is widespread and influential in the philosophy of science (especially in the form of Bayesian confirmation theory). De Finetti was one of the three independent thinkers (James Ramsey in the USA, Émile Borel in France) who defended Bayesian approaches, while Fisher, Egon Pearson, and Neyman lead the then victorious frequentist paradigm. De Fineti's subjective interpretation of probability was part of an ambitious and original approach to the problems of statistical thinking. He contributed equally to methodological as well as to philosophical debates on Bayesianism. For example, he talked about the *coherence* of subjective probabilities (having fair odds that avoid sure loss) and pioneered the concept of "exchangeability," or *exchangeable* random variables (where permutation symmetric subjective probabilities over a sequence of variables may be represented by mixtures of *iid* statistical probabilities, Vicig and Seidenfeld 2012). The truth is that de Finetti was a thinker as well as mathematician, trying to understand the reality that justifies causation as well as the methods that can provide knowledge about events connections. The key problem, for him, as he expressed in *Probabilismo—Saggio critic sulla teoria delle probabilità e sul valore delle scienza* (Perrella: Italy, 1931)—was to show how to connect internal feelings (*sensazione psicologica*, Sect. 24) with fictional quantities, which express a possible

---

[4]According to Sprenger (2012), Bernardo works on something called "reference Bayesian approach", a desubjectivization of the Bayesian account while at the same time maintaining its decision theoretic foundation. The key point here is how to understand the value of loss functions, which should not vary under one-to-one transformations. That is, invariant loss functions. At the same time, it is necessary to select reference priors, something that Bernardo makes maximizing the information of the data, "in maximising the information that the data transmit about the parameter of interest" (ibid., p. 5). Finally, Bernardo's ideas on reference Bayesianism allow a unified approach to hypothesis testing and estimation.

[5]Bradley (2012) suggests a similar approach to de Finetti that he calls "imprecise probabilism." Bradley says that it is a deidealised versión of Bayesianism that represents an agent's belief state by a *set* of probability measures, rather than by a single such measure. Well, it make possible a more complex quantification approach to subjective beliefs, it is true.

[6]There was an extra factor against Bayesians: the accusation of Bayesians of being 'socialists' (during McCarthy's campaign!!!), or un-American. Bertsch (2011: 87).

grading of beliefs.[7] At the end, he thought, we have no direct contact with reality, but with cognitive processes[8] that allow us to infer values from data. For that reason in 1934, he published *The Invention of Truth* (*L'invenzione della verità*), a philosophical text in which he tried to create a philosophical framework for the emergence of a statistical mechanism that can provide knowledge from reality. His idea is that logic does not explain anything and that external reality is a construct that we can imagine thanks to several sets of concepts, such as space, time, matter, or energy (§30, "Ciò che è logico è essato, ma non dice nulla...Per imaginare che cosrrispondono a una "realtà esterna" dobbiamo prima inventare la "realtà esterna" immaginando un modello fisicomatematico (spazio, tempo, material, energia) cin cui rappresentare e esteriorizzare le nostre impressioni"). Then, causes are not real, but they are projections of our mind over external events, and evidence is an illusion. The method is then all.[9] Consequently, truth is probabilistic, and probability is plausible.[10]

At the same time, we must notice that likelihoodism was a well-defined third branch of statistics with no common agreement among statistics experts. In my case, I will include them into conceptually frequentist-oriented experts with a less restrictive view (or realist) on the objectivity issues. Hawthorne (2011: p. 358), defines this "third way" as:

---

[7]Well, this a key point not considered by most of evidence seekers, like Hill and his 9 criteria or Mayo and Spanos (2010), when they affirm that a viable gravity theory must be complete, self-consistent, and relativistic and have the correct Newtonian limit. All these criteria are theoretical and are not accompanied by a quantification mechanism as well as with a rule for not deciding between these variables once quantified. From psychological ideas to absolute truths and without thinking about how these really are connected. It is a naive frequentism that hates Bayesianism because it is supposed that Bayesianism is too subjective. Their project turns absolutely crazy once they defend frequentism as a good inductive inference, when induction is precisely opposite to frequentist ideas, because there is a bias in the fact selection and on the limitation of the long-run experimental design and analysis. Nobody makes infinite experiments! Then, there is an agreement or expert protocol beyond quantified data about the value of the obtained data.

[8]In *Probabilism*, first page, he says that human thought (il pensiero) is a mere biological function (una funzione biologica), a method to orientate one's life (un mezzo per orientarsi nella vita). Besides, sciences cannot provide absolute truths as if they were measuring experimental results: There is always a subjective processing of the data (Sect. 14). Here, Causality must be understood as a mental mechanism necessary for the understanding and control of facts. At Sect. 9, he adds "The probability of an event is relative to our degree of ignorance" (La probabilità di un evento è dunque relativa al nostro grado d'ignoranza). Here, a priori knowledge is illusory (or even proofs that converge toward the infinite) despite the validity of logical and formal tools. He defends, in Sect. 14, the practical truth (certeza pratica), saying that it is not positivist nor rationalist (Sect. 16).

[9]As he defends at (1934) *L'invenzione della verità*: To analyze everything? OK, but with what? (§4. "Analizzare tutto? Va bene. Ma con che?").

[10]In his writings for the Lecture for the Accademia del Lincei, de Finetti wrote in 28/07/1973, that the relationship between mathematics and plausibility belongs to Polya (onine Bruno de Finetti Papers, 1924–2000, ASP.1992.01, Box 6, Folder 10). Anyhow, de Finetti was in disagreement with Polya, as a personal letter of January 7th, 1975 shows, because the latter defended frequency as the value of long-run frequencies.

A view (or family of views) called likelihoodism maintains that confirmation theory should only concern itself with how much the evidence supports one hypothesis over another, and maintains that evidential support should only involve ratios of completely objective likelihoods. When the likelihoods are objective, their ratios provide an objective measure of how strongly the evidence supports h_i as compared to h_j, one that is "untainted" by such subjective elements as prior plausibility considerations. According to likelihoodists, objective likelihood ratios are the only scientifically appropriate way to assess what the evidence says about hypotheses. Likelihoodists need not reject Bayesian confirmation theory altogether. Many are statisticians and logicians who hold that the logical assessment of the evidential impact should be kept separate from other considerations. They often add that the only job of the statistician/logician is to evaluate the objective strength of the evidence. Some concede that the way in which these objective likelihoods should influence the agents' posterior confidence in the truth of a hypothesis may depend on additional considerations — and that perhaps these considerations may be represented by individual subjective prior probabilities for agents in the way Bayesians suggest. But such considerations go beyond the impact of the evidence. So it's not the place of the statistician/logician to compute recommended values of posterior probabilities for the scientific community.

Efron (1998: 111) suggested that the boundaries among statistical schools were not so clear, (see Fig. 5.1):

So, although we consider in this book the existence of two main approaches to statistics, Bayesian and frequentist, we could consider the possibility to include a third one into the list, fisherian. But we also need to admit that Fisher was frequentist except for his fiducial argument, and nobody ascribed themselves to it. It does not mean that really a great divergence exists among statistics practitioners: in

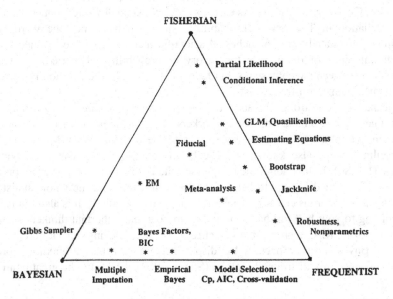

Fig. 5.1 Relationship between different schools of statistics. Taken from Efron (1998: 111). Published with the kind permission of © Institute of Mathematical Statistics, 1998. All rights reserved

the case of theoreticians, perhaps this could be true (a deep distance between Bayesian and frequentists), but from a practical point of view, experts try to use the best tools for each specific problem and this implies not to be married exclusively with a philosophical approach, instead showing an open attitude toward the best rational methods for problem-solving tasks. Before we see contact points among them, as Kass (2011: 1) tried to create a merged approach that he calls "statistical pragmatism," we will analyze briefly some of the biggest debates in this field.

5.2   An Acid and Blurred Debate

Statistics is a young and dynamic discipline and, consequently, opens to strong and serious debates among their practitioners. Some frictions can also be found in the same intellectual side, as happened to Karl Pearson, and Ronald Fisher, but for personal reasons that are not the main interest of this text (Inman 1994). Fisher and Jeffreys had a dispute between 1932 and 1934 because of their opposing views on the nature of statistics, as well as Fisher (1955) with Neyman, Fisher made an intense critique of Neyman ideas, which received a direct answer by Neyman (1956). Fisher even proclaimed "mad Neymanians in California" (Louçã 2008: 6). The disputes were hard and increased once Neyman–Pearson's methods gained general recognition and were adopted swiftly through the academic world. After World War II, it became the received position. Fisher was perhaps the most combative statistician of the twentieth century, entering into public disputes with Neyman, Gosset, and Wishart. As early as in 1949, Kendall wrote a paper, "On the Reconciliation of Theories of Probability," in which he coined the word "frequentist" and stated "Few branches of scientific method have been subject to so much difference of opinion as the theory of probability." He tried to attempt mediation between the contestants, but failed. Curiously, he did mention Bayesians, but named them "non-frequentists."

He had such a strong character that while he was working in 1928 on crop experiments at Rothamsted, some workers sung a funny Christmas song at Rothamsted annual Christmas party (it was to the tune of "Wrap Me Up in My Tarpaulin Jacket," also known as "The Handsome Young Airman Lay Dying") (Kruskal 1980). In that song, they expressed the dominance and power of Fisher, who was presented as able to bend formulae to his will, and the whole statistical activity as a wondrous cult for lay and non-mathematical minds. It is also funny/sad (according to your half full-half empty day) to note that important thinkers such as Karl Pearson, R.A. Fisher, Harold Jeffreys, or Ian Hacking misinterpreted the original Bayes' essay, especially the disputed assumption that an unknown probability is uniformly distributed a priori (Stigler 1982).[11] Despite the great conflicts

---

[11]Stigler argues that although there is a relationship between Bayes argument and the principle of insufficient reason, it is a distant relationship. When Bayes characterized the lack of knowledge

between Fisher and Neyman/Pearson, economists ignored this debate and mixed both approaches, thus creating a methodological hybridization (Louçã 2008), following the term coined by Gigerenzer et al. (1989: 106).

In our days, some Taliban-like approaches to statistical thinking can still be found: For example, a recent keynote at an IACAP conference (2014), Selmer Bringsjord and Naveen Sundar, gave a speech titled "Two refutations of Hegemonic Bayesianism." There, the authors affirm things like "the sign of a mind *infected* with either or both of these views (both Bayesian)" or "dirt-simple Baye's Theorem." The authors, luckily satisfied by their intellectual superiority, even affirmed that the problem of the statistical mechanisms of the human brain was evidently non-Bayesian and that this problem will not be discussed then but "saved for five minutes on another day."[12] Prepotence, lack of knowledge about real scientific practices, and isolation into a blind ivory tower are the only explanations that I can find to justify such an uninformed, weak, and outmoded attitude.

Clearly, one of the recurrent arguments against/in favor of one of the two positions (frequentist or Bayesian) consists in saying that a true scientist is always/never frequentist/Bayesian (you can *choose* between the two possibilities). As an example of this confrontation, see the ideas of Giere (1988): "Are Scientists Bayesian Agents? (...) The overwhelming conclusion is that humans are not Bayesian agents", and of Efron (1986) or Cousins (1995). The last two do not need to be quoted. It seems to be an epistemological law about statistical practices: "A true scientist never belongs to the opposite statistical school" (Vallverdú 2008). It could seem that frequentists are realists, when they consider relative frequencies, and that Bayesians are subjective, when they defend degrees of belief of prior probabilities, but the truth is that in cluster investigations, for example, the frequentist approach is just as subjective as the Bayesian approach,[13] although the

---

(Footnote 11 continued)

about θ by a discrete uniform distribution for *X,* he did not postulate that he was unwilling to specify probability *a priory:* In fact, his basic definition of probability was an a priori expectation (Stigler 1982: 253).

[12]The truth is that brains seem to work in another direction: Synaptic plasticity shows an inherent stochastic nature. Besides, spine mobility enables cortical networks of neurons to carry out probabilistic inferences by sampling from a posterior distribution of network configurations, Kappel (2015).

[13]For example, implicit in the best Bayesian practice is a stance that has much in common with the error statistical approach of Mayo (1996), despite the latter's frequentist orientation (Gelman and Shalizi 2013). They, Gelman and Shalizi (2013) defend that Bayesianism do not support any particular philosophical school but instead fit modern forms of hypothetico-deductivism. Well, I agree with them that actual Bayesian tools do not represent a pure inductivist approach, but at a certain point a sophisticated deductive perspective. Considering the fact that statistical tools are an important but not dominant part of all scientific activities, they can be understood as a bunch of tools that satisfy specific conceptual requirements throughout a long process of epistemic activities. On the other hand, the notion that Bayesian data analysis approaches it to falsificationism (Gelman and Shalizi 2013: 17) must be seen as a naive view about how scientific research works. Scientists are not falsificationists. It is not an explicit step of any of the existing epistemological heuristics, just a possible rule that is optionally applied.

Bayesian approach is less ambitious in that it treats the analysis as a synthesis of data and personal judgments (possibly poor ones), rather than objective reality (Coory et al. 2009).

5.3   Why Bother to Become Frequentist/Bayesian?

Bland and Altman (1998: 1160) have their own answer: "Most statisticians have become Bayesians or Frequentists as a result of their choice of university." And as the epidemiologist, Berger says (2003): "practicing epidemiologists are given little guidance in choosing between these approaches apart from the ideological adherence of mentors, colleagues and editors."

So, the arguments go beyond the ethereal philosophical arena and become more practical ones. Better opportunities to find a good job is an important argument, and the value of Bayesian academic training is now accepted: "where once graduate students doing Bayesian dissertations were advised to try not to look too Bayesian when they went on the job market, now great numbers of graduate students try to include some Bayesian flavor in their dissertations to increase their marketability" Wilson (2003: 372). Therefore, and following Hacking (1972: 133): "Euler at once retorted that this advice is metaphysical, not mathematical. Quite so! The choice of primitive concepts for inference *is* a matter of "metaphysics." The orthodox statistician has made one metaphysical choice and the Bayesian another." To be honest, this is not a fatally flawed position, but different context-based applications of both main approaches. As Gigerenzer (1990) expresses, "we need statistical thinking, not statistical rituals." Lilford and Braunholtz (1996: 604) goes further: "when the situation is less clear cut (...) conventional statistics may drive decision makers into a corner and produce sudden, large changes in prescribing. The problem does not lie with any of the individual decision makers, but with the very philosophical basis of scientific inference. We propose that conventional statistics should not be used in such cases and that the Bayesian approach is both epistemologically and practically superior."[14] There is also a structural aspect: computational facilities; due to recent innovations in scientific computing (faster computer processors) and drastic drops in the cost of computers, the number of statisticians trained in Bayesian methodology has increased (Tan 2001). Trying to offer a mid-point perspective, Berger (2003) proposes using both models and studying their possibilities case by case: "based on the philosophical foundations of the approaches, Bayesian models are best suited to addressing hypotheses, conjectures, or public-policy goals, while the frequentist approach is best suited to those epidemiological studies which can be considered 'experiments', i.e. testing constructed sets of data." Usually, we find no such equitable position.

---

[14]See the persuasive updated from 1997: Hajek (2009).

In the middle of such debates, sometimes really intense and aggressive, there is a space for humor and irony. For example, there is a superb paper from Cohen "The Earth is Round ($p < .05$)," published in 1994, of which the main aim is to rant against the misuse of p-values for null hypothesis significance testing (NHST). The author uses the word of "the ritual of null hypothesis significance testing" or "a sacred .05 criterion" to justify the necessity to abandon faith in specific statistical methods and to embrace the need of experimental replication. The most amusing point of this paper is that it demonstrates that certain Bayesian uses of inverse probability are as false or misleading as frequentist p-values. In a similar joking but philosophically justified spirit, we find the paper of Gigerenzer (1993). The misinterpretation of NHST and H_o affected researchers such as Guilford, Nunnally, Anasti, Ferguson, or Lindquist, besides the error of R.A. Fisher who rejected Bayesian theory of inverse probability but slipped into invalid Bayesian interpretation of NHST (Cohen 1994: 999).

It must be noticed that although far from a more centered perspective critical with both approaches, we can find the delicious Bayesian songs sung at Valencia international conferences[15] where have performed very successfully a choir of female statisticians/singers called "The Bayesettes." A reading of the texts of these musical versions[16] is unavoidable to understand with a smile on your face about the nature of the debate between Bayesians and frequentists. The first international conference on Bayes rule was held in a Spanish city, Valencia, in 1979, and since then, several other Bayesian conferences followed the same location (Bernardo 1999).

5.4 A Practical Mixture and Mutual Influence

Once you have read previous sections of this book, the reader could think that there is a strong battle among statistics experts clearly located at two opposite sides: Bayesians versus frequentists. If it is true that a dichotomization of both schools exists, and that sometimes strong divergences exist about the interpretation of statistical tools, it can be affirmed at the same time that once we take into account practical frameworks in which there are needed results, boundaries are blurred by common sense. For example, there are frequentists as well as Bayesian parametric, semi-parametric, and nonparametric models with specific tools for similar problems that can be combined. And there is an important parallel between frequentist and Bayesian methods: their dependence on the model chosen for the data probability

---

[15]You can easily find videos of these amateur performances at: https://www.youtube.com/watch?v=zO63SQkeuW0 or https://www.youtube.com/watch?v=RRhIxf6EIDg.

[16]The Valencia songbook can be downloaded from: https://www.google.es/url?sa=t&rct=j&q=&esrc=s&source=web&cd=2&ved=0CDoQFjAB&url=http%3A%2F%2Fwww.biostat.umn.edu%2F~brad%2Fsongbook.ps&ei=gqUkUvWENoKL7AbLj4HIDw&usg=AFQjCNEzNlZru33lEtj-IImj2Juu1RLBig&sig2=dYfR8P_M6wZR_2SJoqCCsA.

p (data|parameters) (Greenland 2006: 767). Then, frequency-based priors should be called "empirical" rather than "subjective."

There is another possibility: a pragmatic approach to statistics, something that we cannot forget or neglect. Even in the case of one of the contemporary founders of frequentism, Pearson, we can find a good attitude toward the use of Bayesian methods in specific contexts. In 1917, Pearson and other authors (headed by Soper) wrote *On the distribution of the correlation coefficient in small samples. Appendix II to the papers of "Student" and R.A. Fisher. A cooperative study* (H.E. Soper, A.W. Young, B.M. Cave, A. Lee, and K. Pearson), in which they explained: "Clearly the only result of experience by which we could justify this 'equal distribution of ignorance' would be the accumulative experience that in past series the correlation of parent and child had taken with equal frequency of occurrence every value from −1 to +1. To appeal to such a result is absurd; Bayes' Theorem ought only to be used where we have in past experience, as or example in the case of probabilities and other statistical ratios, met with every admissible value with roughly equal frequency" (p. 358). Howie (2002) claimed that Pearson connected this pragmatic view with the principle of insufficient reason (p. 59).

Again considering the notion of pragmatism, we can read Wilson (2003): "their methodological successes (from Bayesian) have indeed impressed many within the field and without, but those who have adopted the Bayesian methods have often done so without adopting the Bayesian philosophy." As the editorial from British Medical Journal (1996) states, "most people find Bayesian probability much more akin to their own thought processes. The areas in which there is most resistance to Bayesian methods are those where the Frequentist paradigm took root in the 1940s–1960s, namely clinical trials and epidemiology. Resistance is less strong in areas where formal inference is not so important, for example, during phase I and II trials, which are concerned mainly with safety and dose finding." The Scientific Information and Computing Center at CIBA-GEIGY's Swiss headquarters in Basle moved toward the systematic use of Bayesian methods not so much as a result of theoretical conviction derived from philosophical debates, but rather as a pragmatic response to the often experienced inadequacy of traditional approaches to deal with the problems with which CIBA-GEIGY statisticians were routinely confronted (Racine et al. 1986). As an example, clinical trials made by pharmaceutical industries are usually Bayesian method (Estey and Thall 2003) although such methods are not easily implemented (Wang et al. 2002).

Bayarri and Berger paper of 2004 is a perfect example of this pragmatic approach: Going beyond the "philosophical correctness" of any statistical school,[17]

---

[17]Bayarri and Berger (2004), both true Bayesians, remark in p. 60, that "there is a sense in which everyone should ascribe to frequentism: FREQUENTIST PRINCIPLE. In repeated practical use of a statistical procedure, the long-run average actual accuracy should be no less than (and ideally should equal) the long-run average reported accuracy". What they try to show is that some parts of frequentist approach are good and necessary, unless you assume too strictly with their principles.

there are several situations in which a joint frequentist–Bayesian approach is desirable[18], such as

- Mixed approaches:
  - Empirical Bayes approach, by Robbins (1955, 1964, 1983) and other modern reviews of this (Carlin and Louis 2000; Robert 2001).
  - Gamma minimax approach (Berger 1985; Vidakovic 2000).
  - Restricted risk Bayes (Berger 1985).
  - Prequential approach (Dawid and Vovk 1999).
  - Binary regression: the BARS-based (Bayesian adaptive regression splines) posterior interval can be considered from either a Bayesian or frequentist point of view (Kass 2011: 3).

- Design or preposterior analysis: frequentist design focused on planning of experiments is very close to the Bayesian notion or preposterior analysis.
- Binomial confidence interval.
- Estimation and confidence intervals: frequentist estimates and confidence intervals coincide exactly with the standard objective Bayesian estimates and credible intervals.
- Subjectivism: according to Louçã (2008), Fisher arbitrarily chose the model and the test; Neyman and Pearson made the same with the class of hypotheses and the rejection region, and finally, Bayesian had the a priori probability. A last example provides evidence that α-levels are arbitrary (why .05 instead of .06?). Obviously, this needs mode details, but in essence, it is quite correct, and the good attitude toward several grades of subjectivism among statistical studies must be respected and properly understood.
- Computation with hierarchical, multilevel, or mixed model analysis, such as Gibbs sampling and other Markov chain Monte Carlo (MCMC) methods of analysis. MCMC methods rely fundamentally on frequentist reasoning to do computation.
- Assessment of accuracy of estimation: finding good confidence intervals in the presence of nuisance parameters, obtaining good conditional measures of accuracy, and making accuracy assessments in hierarchical models.
- Foundational level: to find the complete class of procedures optimal to solve a situation, frequentists and Bayesians use similar approaches. Frequentist use of minimaxity (Brown 1994, 2000; Strawderman 2000) is close to Bayesian ways to the "least favorable prior distribution." And Bayesian exchangeability, as developed by de Finetti (1970), involves frequentist reasoning or mathematics by considering infinite series of observations.
- Likelihood function: Both classical and Bayesian statisticians can agree about the importance of the likelihood function, in particular on its value as a concise summary of the experimental data (Cousins 1995).

---

[18]Williamson (2013) defends the same position, from a more philosophical rather than an operational perspective. A note published but officially uploaded paper with ideas of Thomas J. Loredo is highly recommendable: *"The Return of the Prodigal: Bayesian Inference in Astrophysics."*

- Prior development: frequentist consistency principle.
- Frequentist simplifications and asymptotic approximations.
- p-Values, for both classic frequentist and Bayesian statisticians (Greenland and Poole 2013). It can even be said that fisherian p-values were incoherently mixed with Neyman–Pearson's hypothesis tests and confidence intervals (Greenland 2006: 765).
- Bayesian networks,[19] despite the name, Bayesian networks do not necessarily imply a commitment to Bayesian statistics.[20] Indeed, it is common to use frequentist methods to estimate the parameters of the conditional probability distributions (CPDs). Rather, they are so-called because they use Bayes' rule for probabilistic inference, and the term "directed graphical model" would be perhaps more appropriate. Nevertheless, Bayes nets are a useful representation for hierarchical Bayesian models, which formed the foundation of applied Bayesian statistics. In such a model, the parameters are treated like any other random variable and become nodes in the graph.
- Bounded rationality and experts' heuristics: Although it is clear that normal human beings do not follow rules of logic, probability theory, statistics, rational choice, or game theory during their daily decisions, experts use some of these tools when they work in their fields. Several fallacies and biases have been discovered during the last decades that affect how experts process numerical data (overconfidence bias, Alexander Problem, Linda Problem, Harvard Medical School Problem, Violation of expected utility theory, Gambler's fallacy, etc.), but at the same time, the mixture of Bayesian and frequentist approaches, or more precisely, its selective implementation, can contribute to controlling these mistakes. As Thomas et al. (2009) have shown the probability to give correct answers under uncertainty decreases with the level of education, some procedures are biased because of the wrong combination of new information with prior information as well as because of the difficulties of making judgments that involve conditional probabilities. The base rate neglect is a very common bias, and for some wrong heuristics like the Harvard Medical School Problem (Casscells and Schoenberg 1978), there is a solution: use different number treatment tool: from Bayesian to frequentist. At that moment, base rate fallacies drop to 30 % and less (Cosmides and Tooby 1996; Gigerenzer and Hoffrage 1995). Then, we could apply Bayesian methods in order to obtain more accurate and reliable data but process them cognitively under a frequentist framework to improve the whole heuristics or involved experts. If classic Bayesian efforts toward solving the

---

[19]Quoted from: Murphy (1998).

[20]As Gillies and Sudbury (2013) show, Bayesian networks can often be improved by interpreting the probabilities objectively and testing the so-called Markov assumption, using some of the frequentist statistical tests. The beneficial symbiosis between both paradigms is evidence!

problem of induction with the help of probability calculus yield to cognitive mistakes (Albert 2009), we can design *new methods of data visualization* or even to combine methodologies to provide a better final outcome.[21]

Nevertheless, there still remain some deep gaps between both schools such as multiple comparisons, sequential analysis, or finite population sampling. Besides, some narrowness on the interpretation of the epistemological value of certain statistical tools, such as happens usually with *p*-values plus a generalized lack of replication, has led to some researchers to distrust main research results, like Ioannidis (2005) and his shocking paper titled: "Why Most Published Research Findings Are False," in which he also affirms: "(...) large-scale evidence is impossible to obtain for all of the millions and trillions of research questions posed in current research" (p. 700). Perhaps one of the most unnoticed but deep problems of frequentist techniques has been the strict perspective about the range of available statistic tools as well as sometimes naïve perspective about how real science is performed, something broadly extended during the first half of the twentieth century by logic positivists. Then, when Bayesian methods were reintroduced, they spread fast among scientists, not for philosophical reasons but for practical ones.[22] Despite Savage's success in 1954, giving to the subjective interpretation of probability a solid philosophical basis, in my opinion, challenges several researchers were faced with were solved in the next decades, and this is what gave Bayesianism the real support. Most times, actions determine the success of theories.

References

Albert, M. (2009). Why Bayesian rationality is empty, perfect rationality doesn't exist, ecological rationality is too simple, and critical rationality does the job. In M. Baurmann & B. Lahno (Eds.), *Rationality, markets and morals: Perspectives in moral science* (pp. 29–65). Frankfurt, Main: Frankfurt-School-Verlag.
Bayarri, M. J., & Berger, J. O. (2004). The interplay of Bayesian and frequentist analysis. *Statistical Science, 19*(1), 58–80.
Bayes, T. (1763). An essay towards solving a problem in the doctrine of chances. *Philosophical Transactions of the Royal Society of London, 53*, 370–418.
Berger, J. (1985). *Statistical decision theory and Bayesian analysis*. NY: Springer.
Berger, Z. D. (2003). Bayesian and frequentist models: Legitimate choices for different purposes. *AEP, 13*(8), 583.
Bernardo, J. M. (1999). The Valencia story. *ISBA Newsletter, 6*, 7–11.

---

[21]About data visualization and new statistical tools, you should look carefully at: http://www.gapminder.org/, accessed April, 21, 2014. As they describe themselves "The mission of Gapminder Foundation is to fight devastating ignorance with a fact-based worldview that everyone can understand. We started the Ignorance Project to investigate what the public know and don't know about basic global patterns and macro-trends. We use surveys to ask representative groups of people simple questions about key-aspects of global development."

[22]See for instance, Kim (2005) paper: "Bayesian dual threshold design with Dirichlet distribution: An alternative way to frequentist multi-stage phase II designs in oncology".

Bernardo, J. M., & Smith, A. F. M. (1996). *Bayesian theory*. USA: Wiley.

Bertsch McGrayne, S. B. (2011). *The theory that would not die: How Bayes' rule cracked the enigma code, hunted down Russian submarines, and emerged triumphant from two centuries of controversy*. USA: Yale University Press.

Bland, M. J., & Altman, D. G. (1998). Bayesian and frequentists. *British Medical Journal, 317*, 1151–1160.

Bradley, S. (2012). *Scientific uncertainty and decision making*. Ph.D. Thesis, London: LSE.

Brown, L. D. (1994). Minimaxity, more or less. In S. Gupta & J. Berger (Eds.), *Statistical decision theory and related topics* (pp. 1–18). NY: Springer.

Brown, L. D. (2000). An essay on statistical decision theory. *Journal of American Statistical Association, 95*, 1277–1281.

Carlin, B., & Louis, T. A. (2000). *Bayes and empirical-Bayes methods of data analysis*. NY: Chapman and Hall.

Carnap, R. (1950). *Logical foundations of probability*. Chicago: University of Chicago Press.

Casscells, W., Schoenberger, A., & Graboys, T. (1978). Interpretation by physicians of clinical laboratory results. *New England Journal of Medicine, 299*(18), 999–1001.

Cohen, J. (1994). The earth is round ($p < .05$). *American Psychologist, 49*(12), 997–1003.

Coory, M. D., Wills, R. A., & Barnett, A. G. (2009). Bayesian versus frequentist statistical inference for investigating a one-off cancer cluster reported to a health department. *BMC Medical Research Methodology, 9*, 3.

Cosmides, L., & Tooby, J. (1996). Are humans good intuitive statisticians after all? Rethinking some conclusions from the literature on judgment under uncertainty. *Cognition, 58*(1), 1–73.

Cousins, R. D. (1995). Why isn't every physicist a Bayesian? *American Journal of Physics, 63*(5), 198–410.

Dawid, A. P., & Vovk, V. G. (1999). Prequential probability: Principles and properties. *Bernoulli, 5*(1), 125–162.

de Finetti, B. (1931). *Probabilismo – Saggio critic sulla teoria delle probabilità e sul valore delle scienza*. Italy: Perrella.

de Finetti, B. (1937). *La prevision: ses lois logiques, ses sources subjectives*, Ann. Inst. H. Poincaré, 7: 1–68. (English translation in H. E. Kyburg & H. E. Smokler (Eds.) (1964) *Studies in subjective probability*). New York: Wiley.

de Finetti, B. (1970). *Teoria delle* Probabilità 1, 2, Torino: Einaudi. (English translations published (1974, 1975) as *Theory of probability 1, 2*. New York: Wiley).

Editorial. (1996). Bayesian statistical methods: A natural way to assess clinical evidence. *British Medical Journal, 313*, 569–570.

Edwin, T., & Jaynes, E. T. (1968). Prior probabilities. *IEEE Transactions on Systems Science and Cybernetics, 4*(3), 227–241.

Efron, B. (1986). Why isn't everyone a Bayesian? *American Statistician, 40*, 1–5.

Efron, B. (1998). R.A. Fisher in the 21st century. *Statistical Science, 13*(2), 95–122.

Estey, E. H., & Thall, P. F. (2003). New designs for phase 2 clinical trials. *Blood, 102*(2), 442–448.

Fisher, R. (1955). Statistical methods and scientific induction. *Journal of the Royal Statistical Society, Series B, 17*, 69–78.

Fitelson, B. (2001a). A Bayesian account of independent evidence with applications. *Philosophy of Science, 68*(3, supplement), S123–S140.

Fitelson, B. (2001b) *Studies in Bayesian confirmation theory*. Ph.D. thesis, University of Wisconsin–Madison (Philosophy). The thesis can be downloaded from http://fitelson.org/thesis.pdf.

Gelman, A., & Shalizi, C. R. (2013). Philosophy and the practice of Bayesian statistics. *British Journal of Mathematical and Statistical Psychology, 66*(1), 8–38.

Giere, R. (1988). *Understanding scientific reasoning* (p. 189). USA: The University of Chicago.

Gigerenzer, G. (1993). The supergo, the ego and the id in statistical reasoning. In G. Keren & C. Lewis (Eds.), *A handbook for data analysis in the behavioral sciences: Methodological issues* (pp. 311–339). Hillsdale, NJ: Erlbaum.

Gigerenzer, G., & Swijtink, Z. (1990). *The empire of chance: How probability changed science and everyday life.* Cambridge: CUP.

Gigerenzer, G., & Hoffrage, U. (1995). How to improve Bayesian reasoning without instructions: Frequency formats. *Psychological Review, 102*(2), 684–704.

Gigerenzer, G., et al. (1989). *The empire of chance: How probability changed science and everyday life.* UK: Cambridge University Press.

Gillies, D., & Sudbury, A. (2013). Should causal models always be Markovian? The case of multi-causal forks in medicine. *European Journal for Philosophy of Science, 64*(1), 275–308.

Good, I. J. (1971). 46.656 kinds of Bayesians. *American Statistician, 25,* 62–63.

Greenland, S. (2006). Bayesian perspectives for epidemiological research I: Foundations and basic methods. *International Journal of Epidemiology, 35,* 765–775.

Greenland, S., & Poole, C. (2013). Living with *p*-values: Resurrecting a Bayesian perspective on frequentist statistics. *Epidemiology, 24*(1), 62–68.

Hacking, I. (1972). Likelihood. *British Journal of Philosophy of Science, 23,* 132–137.

Hajek, A. (2009). Fifteen arguments against hypothetical frequentism. *Erkenntniss, 70,* 211–235.

Hawthorne, J. (2011). Confirmation theory. In D. M. Gabbay, P. Thagard & J. Woods (Eds.), *Handbook of the philosophy: Philosophy of statistics* (Vol. 7, pp. 333–389). Oxford (UK): Elsevier.

Howie, D. (2002). *Interpreting probability: Controversies and developments in the early twentieth century.* Oxford: OUP.

Howson, C. (1997). On Chihara's the Howson-Urbach proofs of Bayesian principles. *British Journal for the Philosophy of Science, 48*(1), 83–90.

Howson, C., & Urbach, P. (1991). Bayesian reasoning in science. *Nature, 350,* 371–374.

Inman, H. F. (1994). Karl Pearson and R.A. Fisher on statistical tests: A 1935 exchange from nature. *The American Statistician, 48*(1), 2–11.

Ioannidis, J. P. A. (2005). Why most published research findings are false. *PLoS Medicine, 2*(8), e124. doi:10.1371/journal.pmed.0020124.

Jeffreys, H. (1939). *Theory of probability.* Oxford: Clarendon Press.

Joyce, J. M. (1992). *The axiomatic foundations of Bayesian decision theory.* Ph.D. Dissertation, University of Michigan.

Kappel, D. et al. (2015). Network plasticity as Bayesian inference. arXiv:1504.05143.

Kass, R. E. (2011). Statistical inference: The big picture. *Statistical Science, 26*(1), 1–9.

Kendall, M. G. (1949). On the reconciliation of theories of probability. *Biometrika, 36,* 101–116.

Keynes, J. M. (1921). *A treatise on probability.* London: Macmillan.

Kim, H. W. (2005). *Bayesian dual threshold design with Dirichlet distribution: An alternative way to frequentist multi-stage phase II designs in oncology.* Texas Medical Center Dissertations (via ProQuest). Paper AAI3182105. http://digitalcommons.library.tmc.edu/dissertations/AAI3182105.

Kruskal, W. (1980). The significance of Fisher: A review of R.A. Fisher: The life of a scientist by Joan Fisher box. *Journal of the American Statistical Association, 75*(372), 1019–1030.

Laplace, P.-S. (1814a). *Essai philosophique sur les probabilités,* Paris: Courcier. References to this work have been made from (1886) *Ouvres complètes de Laplace,* Vol. 7, Paris: Gauthier-Villars.

Laplace, P.-S. (1814b). *Théorie analytique des probabilités, Œuvres complètes, tome 7,* ii-cliii+1–645. Downloadable at http://math-doc.ujf-grenoble.fr/cgi-bin/oeitem?id=OE_LAPLACE__7_R2_0

Lewis, D. (1980). A subjectivist's guide to objective chance. In R. C. Jeffrey (Ed.), *Studies in inductive logic and probability* (Vol. 2). Berkeley, CA: University of California Press.

Lilford, R. J., & Braunholtz, D. (1996). For debate: The statistical basis of public policy: A paradigm shift is overdue. *British Medical Journal, 313,* 603–607.

Louçã, F. (2008). The widest cleft in statistics—How and why Fisher opposed Neyman and Pearson. *Working Papers WP 02/2008/DE/UECE.* Lisbon: ISEG.

Mayo, D. (1996). *Error and the growth of experimental knowledge.* Chicago: U. Of Chicago Press.

Mayo, D. G., & Spanos, A. (2010). *Error and inference: Recent exchanges on experimental reasoning, reliability, and the objectivity and rationality of science*. Cambridge: CUP.

Murphy, K. (1998). A brief introduction to graphical models and Bayesian networks. http://www.cs.ubc.ca/~murphyk/Bayes/bnintro.html. Accessed on November 15, 2013.

Neyman, J. (1956). Note on an article by Sir Ronald Fisher. *Journal of the Royal Statistical Society, Series B, 18*(2), 288–294.

Racine, A., et al. (1986). Bayesian methods in practice: Experiences in the pharmaceutical industry. *Applied Statistics, 35*(2), 93–150.

Ramsey, F. P. (1931). Truth and probability. In P. K. Trench (Ed.), *The foundations of mathematics and other logical essays*. London: Truber.

Robbins, H. (1955). An empirical Bayes approach to statistics. *Proceedings of the Third Berkeley Symposium on Mathematical Statistics and Probability, 1*, 157–164.

Robbins, H. (1964). The empirical Bayes approach to statistical problems. *Annals of Mathematical Statistics, 35*, 1–20.

Robbins, H. (1983). Some thoughts on empirical Bayes estimation. *Annals of Statistics, 11*, 713–723.

Robert, C. P. (2001). *The Bayesian choice: From decision-theoretic foundations to computational implementation*. NY: Springer.

Savage, L. J. (1954). *Foundations of statistics*. NY: Wiley.

Skyrms, B. (1984). *Pragmatics and empiricism*. USA: Yale University Press.

Sprenger, J. (2012). The renegade subjectivist: José Bernardo's reference Bayesianism. *RMM, 3*, 1–13.

Stigler, S. M. (1982). Thomas Bayes's Bayesian inference. *Journal of the Royal Statistical Society, Series A, 145*(2), 250–258.

Strawderman, W. (2000). Minimaxity. *Journal of American Statistical Association, 95*, 1364–1368.

Tan, S. B. (2001). Bayesian methods for medical research. *Annals Academy of Medicine, 30*(4), 444–446.

Thomas, D. et al. (2009). *The non-use of Bayes rule: Representative evidence on bounded rationality*-ROA-RM-2009/1. The Netherlands: Maastricht University. Downloadable at http://edocs.ub.unimaas.nl/loader/file.asp?id=1383. Accessed on April 15, 2014.

Vallverdú, J. (2008). The false dilemma: Bayesian *versus* Frequentist. *E-L O G O S Electronic Journal for Philosophy/2008*, 1–17.

Vicig, P., & Seidenfeld, T. (2012). Bruno de Finetti and imprecision: Imprecise probability does not exist! *International Journal of Approximate Reasoning, 53*(8), 1115–1123.

Vidakovic, B. (2000). Gamma-minimax: A paradigm for conservative robust Bayesians. In *Robust Bayesian analysis, Lecture notes in statistics* (Vol. 152, pp. 241–259). NY: Springer.

Wang, J., et al. (2002). An approximate Bayesian risk-analysis for the gastro- intestinal safety of ibuprofen. *Pharmacoepidemiology and Drug Safety, 11*, 695–701.

Williamson, J. (2013). Why frequentists and Bayesians need each other. *Erkenntnis, 78*(2), 293–318.

Wilson, G. (2003). Tides of change: Is Bayesianism the new paradigm in statistics? *Journal of Statistical Planning and Inference, 113*, 3171–3174.

Chapter 6
The Birth of Multicausality as the Death of Causality and Their Statistical Corollaries

Abstract The emergence of a new discipline, epidemiology, contributes to the understanding of the evolution of scientific attitudes and ideas about causality. After an initial and ancient belief in single causes supported by classic philosophers and nineteenth century physicians like Koch that can be expressed as a monocausality view, the complexity of real medical and toxicological problems forced researchers to embrace the notion of multicausality and similar approaches (web of causes, chain of events). All these debates fed the evolution of statistical methodologies employed as well as led to a new way to understand causality within complex systems or contexts.

Keywords Epidemiology · Monocausality · Multicausality · Risk · Web of causes · Chain of causal relationships · Evidence · Social · Complexity

Causality debates did not finish with the evolution of statistical tools and instruments. In fact, the more we know about the world, the more we need to admit our ignorance and oversimplification of data we extract from nature. In this chapter, I will introduce the last important debate on induction, causality and statistics: the epidemiological debate.

In this chapter we will analyze a very important topic: the multiple hypotheses problem. If historically philosophers have been worried about the notion of cause, it was a mono-cause problem: how one thing causes another thing. It was a single cause dogma, absolutely tacit for all the academic community. But with the progress of sciences, the notion of multiple causes or cocauses emerged as a true force of nature, beyond the reluctances of the scientists and theoreticians and impelled by the new complexities of modern sciences (immersed today in Big Data problems).

Very briefly let us take an eagle-eye view on the historical process that preceded the birth of multicausality: the bygone era of single causes and absolute truths. In the Western philosophical tradition, the search for non supranatural causes to explain the world has been the fundamental force of research, despite all the exceptions and counterexamples we could find to this affirmation: one event, one cause, and the human mind (with the help or not of sensory data) ability to catch it. But nature seems to behave in different ways at different scales or reality domains,

J. Vallverdú, *Bayesians Versus Frequentists*,
SpringerBriefs in Statistics, DOI 10.1007/978-3-662-48638-2_6

so it seemed necessary to find a more technical theory of causation and Aristotle set the theory of 4 causes to explain how the world worked, becoming the first Grand Unified Theory, if you allow me the use of contemporary vocabulary. Although he included the notions of chance and causality into his writings (*Physics* II, 4–6), the dogma was: *one effect requires a single cause*. Consequently, the origin of the universe resided into one primordial and temporarily initial cause: the Unmoved Mover, from which all the entities of our universe were a mere consequence. In this universe, everything followed strict rules that made establishing links between causes and effects possible, completely open to the mind of philosophers. Only Lucretius and Democritus were too brave or insensate as to offer a place to hazard or indeterminacy when they considered that atoms followed a basic deviation or *clinamen* to produce movement and interaction. That is, the cause of change. Later, was constructed inside Western thought an increasingly materialistic and mechanistic approach that made possible one unexpected consequence: the quantification of several domains of human made possible to assume that, first, human life was quantifiable, and second, that those big numbers needed specific treatment. This opened the backdoor of rationalists to accept the idea of numeric probability, first in the work *La logique ou l'art de penser*, written in 1662 by Antoine Arnauld and Pierre Nicole. Later, other authors made new advances: Huygens, 1657, *De ratiotiniis in aleae ludo*; De Moivre, 1711, *De mensura sortis* (in which the view was defended that hazard was not possible, otherwise it could be a demonstration of atheism's validity)[1]; Laplace, 1789, *Essai philosophique sur les probabilités*; Bayes 1763, *An Essay towards solving a Problem in the Doctrine of Chances*. Some authors understood how this implied the fracture of systematic thinking, like Leibniz, who argued that hazard was not a reality but the result of our ignorance about the system as a whole.

The ethical aspects of indeterminacy were grounded by Pascal and his famous moral bet about the existence of God and our beliefs. How could humans be free and responsible for their acts if the whole world was a deterministic system, a big clock-machine? Some protestant sects were true and we were all damned or forgiven by God before our birth? Descartes tried to find a solution creating a clear distinction between a res extensa forced to follow mechanistic rules and a free *res cogitans* with choice possibility. To us this sounds like a *deus ex machina* fallacy.

Anyhow, the continuous addition of new data to the scientific arena showed that the classic or ancient philosophers and scientists were not the guardians of eternal truths but old walls to be destroyed in order to see the reality. But, alas, this implied at the same time a confusing discovery: New data were not so easy to reduce to lineal causal processes in which A was clearly cause of B. Therefore, causality

---

[1]This was something important for Moivre and more so after his long stays in French prisons as a consequence of his being a Protestant. The battles of God were then performed in a mathematical arena.

became a more complex process, a new challenge for philosophers and scientists. If it is true that statistical thinking emerged as the leading approach in order to solve those problems, a redoubt of classical thinkers tried to maintain a philosophical view. John Stuart Mill was one of them (Auguste Comte was another leading author in this view) and he designed in 1843 (*A System of Logic*) a new way to establish relations of causal necessity among events, five different formal methods for establishing causal connections between types of events: (a) method of agreement; (b) method of difference; (c) joint method of agreement and difference; (d) method of concomitant variations; and (e) method of residues. With these ways to obtain "causal" connections, Mill tried to avoid the type of statistical thinking that was gaining power in scientific and social sciences. Mill denounced statistics as "an aberration of the intellect" and "ignorance...coined into science" (McGrayne 2011: 36). The French statistician Bertrand wrote in 1889: "L'application du calcul aux decisions judiciaries est. dit Stuart Mill, le scandale des Mathématiques. L'Accusation est injuste" (p. 43). Cartwright (1989: 176) asserts that Mill was "opposed" to the statistical laws of Quetelet's social physics, considering necessary the establishment of rules of association of events. This philosophical change was not an easy or desired decision: It was meant the only way to save rationality from chaos and its disappearance. Because, as Bertrand Arthur William Russell, 3rd Earl Russell, affirmed at the beginning of the twentieth century: "the law of causality (...) is a relic of a bygone age, surviving, like the monarchy, only because it is erroneously supposed to do no harm (...) The principle 'same cause, same effect', which philosophers imagine to be vital to science, is therefore utterly otiose" (Suppes 1970: 5). Physicist James Clerk Maxwell also adopted frequency-based methods for the creation of statistical mechanics and the kinetic theory of gases. Statistics was then at the core of nature, not of (limited) minds. Some decades later century Karl Popper wrote *A World of Propensities*, defending a new status for statistical propensities,[2] now becoming the maximum level of rationality.[3] And, later, Nancy Cartwright constructed a defense of probabilistic probability. It looks like the abandonment of philosophical belief in the cause–effect relationships, but it represented an advance into several notions of causality: Entities in the world behave in a more complex way than our predecessors knew, and although we can measure and quantify the whole process, numbers alone are not able to offer direct causal relationships (Buck 1975). In this model, mechanisms replace causes. In fact, scientific disciplines entered into this new approach by operational and experimental results, not by theoretical debates. There is one discipline that exemplifies perfectly this process: *epidemiology*.

---

[2]Philosopher Gillies (2000) defended long-run propensity theories, while his predecessor Popper showed the coexistence of long-run and single-case propensity theories. This is a diffused way to defend frequentist ideas.

[3]About propensities and causes in Quantum Physics see Suárez (2011).

6.1   Epidemology as a Data-Scanner Discipline

According to Bhopal (2002: 21), epidemiology "is the science and craft that studies the pattern of diseases (and health, though usually indirectly) in populations to help understand both their causes and the burden they impose. This information is applied to prevent, control or manage the problems under study." The concept is derived from the Greek words meaning study upon populations (*epi* upon, *demos* people, *logos* study).

The ancient Greek physician Hippocrates (470–400 B.C.) is considered the father of modern Western medicine. He wrote a book entitled *Air, Waters and Places*, where he clearly identified the general dependence of health not on magical and supranatural influences but on an identifiable array of natural external factors, some of them clearly environmental. This led to several historians to talk about the man-environment binominal also referred to as "Hippocrates' dyad." According to Saracci (in Olsen et al. 2010: Chap. 1), there were three streams in early epidemiology: medical, demographic, and theoretical, coalesced in an effective way only toward the end of the eighteenth and beginning of the nineteenth centuries, giving rise to epidemiology as we recognize it today, an investigation of diseases and their etiology at the population level. Ramazzini, Sydenham, Lancisi, or Graunt are some of the pioneers.

A change in the discipline was experienced with the advent of the industrial transformation of Western Europe, when "crowd diseases" emerged which struck the populations amassed in the slums of the fast-growing centers of industrial development. John Snow (1813–1858) exemplifies this renewal of ideas and techniques with his brilliant identification of a pathogenic agent from the environment: the cholera in London because of contaminated water supplies. He contributed to the end of London's great cholera epidemic episodes in 1849 and 1854. Rudof Virchow (1821–1902, talking about medicine as a social science) and Robert Koch (1843–1910, with his causal postulates) were also key actors for the establishment of conceptual elements of epidemiology. After World War Two, this discipline entered into the reign of clinical and statistical debates, around the controversies about tobacco and health. In the next section, we will cover the conceptual analysis on epidemiology evolution about causality and disease.

6.2   The Epidemiological Monocausality

Once practical roots of epidemiological research have been explained, it is necessary to identify the conceptual debate that was born with and became cosubstantial to this discipline: causal relationships. Epidemiology exemplifies perfectly how conceptual debates and empirical advances are intertwined. In 1844, Robert Koch demonstrated how a causal relationship between a bacterial parasite and illness existed, where 1 agent in 1 ambient caused 1 illness. Besides their demonstration,

Koch needed to establish four postulates that reinforced or demonstrated these monocausal relationships (Croce 1996):

1. The microorganism must be found in abundance in all organisms suffering from the disease, but should not be found in healthy organisms.
2. The microorganism must be isolated from a diseased organism and grown in pure culture.
3. The cultured microorganism should cause disease when introduced into a healthy organism.
4. The microorganism must be reisolated from the inoculated, diseased experimental host and identified as being identical to the original specific causative agent.

This successful as well as promising approach implied that, from hence, any expert in diseases looked only for the single agent that caused one disease. The first known agent to behave in that way was anthrax bacterium (*Bacillus anthracis*). French physicists and pathologists, respectively, Casimir Joseph Davaine and Pierre François Olive Rayer,[4] discovered in 1850 certain microorganisms in diseased and dying sheep. That same year Rayer inoculated sheep with the blood of dead sheep (by what we know as anthrax) and published a short note of this work in the *Bulletin de la Societé de Biologie*. Six years later, F.A. Brauel inoculated sheep, horses, and dogs with blood taken from animals sick with anthrax, demonstrating that the disease could be transmitted to sheep and horses, but not to dogs. In 1863, Davaine published three valuable papers on anthrax and demonstrated that the blood of an animal sick with anthrax is not capable of transmitting the disease to others unless it contains the bacillus. Finally, in 1876, German microbiologist Robert Koch researched the etiology of *Bacillus anthracis* and discovered its ability to produce "resting spores" that could stay alive in the soil for a long period of time to serve as a future source of infection. For these strange causes, Koch received all the credit for this epistemologically distributed process. Nevertheless, a "golden age" of scientific discovery ensued. But monocausality was not as certain as it appeared to be as follows: After a few years, the clinical and laboratory evidence showed that this was not true and that a plethora of possibilities were real. As Inglis (2007) shows, there is no single accepted method to establish a causal relationship between an infective agent and its corresponding infectious disease. Different biomedical disciplines use a patchwork of distinct but overlapping approaches. Now, over a century later, a more rigorous method to test causality has still to be finalized. One contender is a method that uses molecular methods to establish a causal relationship ("molecular Koch's postulates").

Before allowing the discipline to enter into a shock or to be collapsed by their inconsistencies, Sir Austin Bradford Hill worked hard establishing causal criteria in epidemiology. In 1965, he published the famous paper "The environment and

---

[4]Rayer was a very famous and successful physician expert among other things in skin diseases. Davaine attended him unsuccessfully in 1867 when he suffered a deadly stroke.

disease: association or causation?" where he specified 9 criteria that made it possible to affirm causal relationships:

1. *Strength*: A small association does not mean that there is not a causal effect, though the larger the association, the more likely that it is causal.
2. *Consistency*: Consistent findings observed by different persons in different places with different samples strengthen the likelihood of an effect.
3. *Specificity*: Causation is likely if a very specific population at a specific site and disease with no other likely explanation. The more specific an association between a factor and an effect is, the bigger the probability of a causal relationship.
4. *Temporality*: The effect has to occur after the cause (and if there is an expected delay between the cause and expected effect, then the effect must occur after that delay).
5. *Biological gradient*: Greater exposure should generally lead to greater incidence of the effect. However, in some cases, the mere presence of the factor can trigger the effect. In other cases, an inverse proportion is observed: Greater exposure leads to lower incidence.
6. *Plausibility*: A plausible mechanism between cause and effect is helpful (but Hill noted that knowledge of the mechanism is limited by current knowledge).
7. *Coherence*: Coherence between epidemiological and laboratory findings increases the likelihood of an effect. However, Hill noted that "… lack of such [laboratory] evidence cannot nullify the epidemiological effect on associations."
8. *Experiment*: "Occasionally, it is possible to appeal to experimental evidence."
9. *Analogy*: The effect of similar factors may be considered.

With these 9 rules (closer to Mill's approach to the causal associationism than to any other mechanistic determinism approach under the statistical framework), Hill tried to establish[5] a solid set of rules that make the advancement or event possible in the existence of epidemiology as a scientific practice (Phillips and Goodman 2004; Höfler 2005). The most amazing fact about Hill's criteria is that he did not provide algorithmic and quantified methods to implement them: This process relies on the expertise and the epidemiologist's own decision. At a certain point, this turns the criteria as declarations of good will, instead of being solid ways to achieve unquestionable truth. Even useful, their implementation is an *epistemological quagmire*.

Anyhow, the increasing number of variables and elements existing at the same time as causal processes sought a need for a new visualization strategy (Greenland et al. 1999) that makes the work of several cocausal factors easier. Part of these criteria was applied in assessing the causal relation between cigarette smoking and

---

[5]Following and unifying ideas from Hume, Hammond, Yerushalmy, Palmer, Lilienfeld, Sartwell, and the Surgeon General's Advisory Committee as a chain of predecessors. See Blackburn and Labarthe (2012: 1075) for these ideas as well as to understand how the debate was richer and with independent ramifications at European and North American institutions.

lung cancer in the 1964 Report of the Advisory Committee to the US Surgeon General, Smoking and Health (Morabia 1991).[6]

Curiously, Hill affirmed in the same paper that "*I have no wish, nor the skill, to embark upon philosophical discussion of the meaning of 'causation'*. The 'cause' of illness may be immediate and direct; it may be remote and indirect underlying the observed association. But with the aims of occupational and almost synonymous preventive, medicine in mind the decisive question is where the frequency of the undesirable event B will be influenced by a change in the environmental feature A. *How* such a change exerts that influence may call for a great deal of research, However, before deducing 'causation' and taking action we shall not invariably have to sit around awaiting the results of the research. The whole chain may have to be unraveled or a few links may suffice. It will depend upon circumstances." On the one hand, Hill declared his lack of interest in the philosophical debate on causation, while on the other hand, he claimed the need for a new way to obtain causal relationships, that is, making epistemology. From a conceptual perspective, Hills's approach seems very similar to the improvements and changes that were suffered in the field of logics: Initially worked with phrases that constituted arguments (syllogisms), until deeper details were entered into using the predicates inside the phrases and thus leading to predicate logic. But this was the logic of perfect abstract minds, not the logic followed by humans. The analysis of how humans really thought introduced new elements into consideration, allowing the emergence of non-monotonic logics, with more and powerful variables to be taken into account. In a similar way, epidemiologists, who initially really believed in material single causes for single effects, shifted toward a more complex panorama in which causes and effects followed several paths that had to be combined in order to produce relations of mutual sequential necessity. Trying to solve how individuals should face Hill and Doll results (devoted to *populations*), Jerome Cornfield used Bayes' rules to create a solution as well as to defend a scientific approach to multicausality, which faced him directly to debate with the fierce Fisher about relationships between tobacco and cancer.

In 1991, Mervyn Susser, after several years of investigations in this area (Susser 1973), proposed to reduce Hill's list to three elements of causal criteria:

a. *Association.* The exposure and outcome are associated more commonly than would be expected by chance.
b. *Time order.* The exposure can be shown to precede the outcome, that is, X precedes Y.
c. *Direction.* A change in the outcome is a consequence of change in exposure, not the same as directionality in a study, where X leads to Y.

---

[6]Morabia (1991) also shows that here are strong evidences towards the strong similitude between first 6 Hill's criteria and Hume's rules (at Hume's *Treatise of Human Nature*).

Five years later, Susser published more research (Susser 1996a, b, together with his son), and even his son Ezra has published very important ideas on causality on his own in epidemiology, especially in psychiatric epidemiology (Susser 2004).

6.3   The Multiple Causes Paradigm in Epidemiological Studies

Trying to put order after the fiasco of monocausation dogma, if we consider it as the representative model for all epidemiological cases, some authors created their own models of multicausality. This was not only necessary for new techniques but also to use metaphors to think about being able to embrace a new way to obtain knowledge about causality.[7] The necessity of this new metaphor is easy to understand: At its beginnings, epidemiology was faced with mono-infectious agents, "easy" to identify and to fight. But in the second half of the twentieth century, health improvements such as filtered water supplies, sewage systems, vaccines for serious epidemic diseases (typhoid fever, tuberculosis, tetanus), discovery of penicillin, better nutrition, and access to sanitary facilities changed the kind of threats that menaced large populations, bringing about a steady decline of infectious diseases during the first half of the twentieth century (Andersen 2007). Chronic and multifactorial diseases emerged as a new research topic, at a moment in which there were no conceptual tools for diseases such as cancer, coronary artery disease, diabetes, ulcers, or strokes. The need for new tools was imperative and soon innovative ways appeared to deal with these diseases: bias and confounding concepts, case–control or cohort designs.[8] This allowed a new set of chronic studies such as the Framingham Heart Study (1948), embarked on an ambitious project in health research to identify the common factors that contribute to cardiovascular disease by following its development over a long period of time in a large group of participants, the study began in 1948 (conducted by Thomas Royle Dawber) with 5209 adult subjects from Framingham, and is now in its third generation of participants; the famous case–control study of cancer cases from London hospitals (Doll and Hill 1950); or the cohort study of British doctors (Doll and Hill 1956). These first studies made it possible to investigate in depth some health problems for which there was not clear data. Lung cancer was one of them, increasing at concerningly greater rates, and soon became a leading research topic among epidemiologists. The causal bonds between lung cancer and smoking were provided by Doll and Hill (1950) and several other authors, running retrospective case–control studies that used hospital populations. Prospective studies were soon to follow, showing defined causal connections between smoking and suffering lung

---

[7]Krieger (1994: 896).

[8]It is true that all these concepts or tools already existed in the previous century, but it was during this period that they reached maturity and precision. See Vandenbroucke (2002), Vineis (2002).

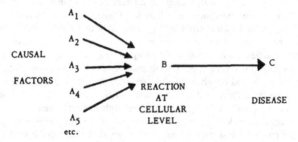

Fig. 6.1 Lilienfled's original image on chain of causal relationships. Taken from (Lilienfeld 1957: 56). Published with the kind permission of © Association of Schools of Public Health, 1957. All rights reserved

cancer, although some leading experts such as the British biostatistician Ronald Fisher or North American biostatistician Joseph Berkson made several objections to these evidences. The weight and influence of Fisher was at that moment over-whelming among epidemiological communities. Not in vain, he was the stronger defender and apostle of randomized clinical trials (RCTs),[9] initially applied to agricultural trials, which he made the gold standard in clinical trials and medical research. According to Fischer, a leading frequentist, RCT avoided biases as well as confounding factors, becoming the only scientific option. Curiously, the first published RCT applied to clinical trials, in 1948 (Marshall et al. 1948), was coauthored by Hill. They used same (or similar) methodological tools, but obtained different or even opposite results.

Back to our analysis of multicausality, we look at the "chain of causation," the first metaphor we will analyze. It emerged after the methodological debates on investigation of chronic diseases (especially cancer and heart diseases) parallel to that of infectious diseases, by leading American biostatisticians, Jacob Yerushalmy and Carroll E. Palmer (Yerushalmy and Palmer 1959). The American epidemiologist Abraham Lilienfeld, surprisingly brother-in-law to Yerushalmy, had described the same assumption that all disease causation would ultimately have to be described at a cellular level (Andersen 2007). This new view he saw as a "chain of causal relationships," a new visual way to understand causality (Lilienfeld 1957: 56) (Fig. 6.1).

Following Lilienfeld own and clear words (1957: 56–59, italics are mine):

> After a *statistical association* has been ascertained, we would like to make some sort of an *inference as to whether a cause and effect relationship exists* between the disease and the associated characteristic (…) Epidemiological studies are composed of two stages: first, the

---

[9]Greenland (2006: 768) affirms that controlled randomized experiments on random samples as a method for scientific research is a mind-projection fallacy, something expressed close to thought experiment completely useless than to real procedures in daily practices. As he affirms: "data alone say nothing at all" (ibid, p. 773).

determination of statistical associations between a disease and various population characteristics; second, the derivation of biological inferences from the pattern of associations.
Both the associations and inferences constitute the epidemiology of the disease. Statistical
associations may be determined from demographic data or from individual history data.
The latter may be obtained from retrospective studies, prospective studies, or experimental
studies. In these studies, characteristics of a group of cases are compared with those of one
or more groups of controls. Cases and controls may be selected by various methods, each of
which has advantages and disadvantages. In general, leads to the existence of statistical
associations come from individual history studies of hospital populations or from demographic data. The associations so suggested require confirmation by retrospective studies of
adequately selected samples of cases and controls from their respective populations.
Whether or not prospective studies are necessary depends largely on the kind and strength
of the association. The method of carrying out a prospective study depends on the nature of
the characteristics and the disease under investigation. *In the derivation of causal inferences
from observed statistical associations, certain biological and non biological factors are
influential.* Among the biological factors are the ability to conduct human experiments, the
strength of the association, the role of animal experimentation, and the prevailing biological
concepts. The latter is time most important. Snow's and Farr's observations on cholera are
illustration-s of the interaction between biological theory and the interpretation of statistical
associations. A non biological factor is the course of action resulting from the degree of
plausibility with which a causal inference is regarded.

"Web of causation" is another of the new successful metaphors and according to
Susser (2001) "is a metaphor for a theory of sequential multiple causes. Originally
deployed mainly for an epidemiology practised at the individual level of organi-
sation, but not necessarily confined to it, the metaphor can be and has been
extended to incorporate a sequence of multiple dimensions." MacMahon et al.
(1960) were the first users of this notion, basically as Krieger (1994) points, "as a
reaction to the prevalent notion of chain of causation." The chain of causation was a
metaphor and research model that failed to account for the complexity of the
genealogy of the antecedent of the chain and, thus, for the possible partial overlap
between different factors. With the web of causation, *multifactorial* etiology of
disease gained ground. This metaphor incited epidemiologists to embrace a more
sophisticated view of causality: that of a spider web made with an intricate network
of delicate and cross-connected strings, symbolizing diverse causal pathways
(Krieger 2011: 151). For example, the *web of causation* of MacMahon et al. (1960)
had a tree structure, close to the family trees we are used to.

6.4   New Trends in Epidemiology and Causality

Once richer, plural, and broader methods of causality were established, new factors
were necessary inclusions into these models.[10] "Eco-social" variables were widely
discussed and multilevel epidemiology went one step ahead of previous webs of

---

[10]It is very curious the close-philosophical conceptualization of this new causality by Rothman and
his causal pies and sufficient causes (Rothman 1976).

causation. In this new paradigm, epidemiological explanations of disease may refer to different levels of analysis of disease causation. Susser and Susser (1996b) talked about the pass from black boxes to Chinese boxes in causal analysis: Then, more levels of complexity do not imply a desertion of causal explanations (Weed 1998), and they show the need to consider the existence of multilevel complex systems. At the same time, several authors avoided a discussion of the notion of cause, because their "goal is to allow researchers to feel more confident in the power of their research to tell a convincing story without resorting to metaphysical/unsupportable notions of cause" (Lipton and Ødegaard 2005: 1). The previous statement is hilarious and terribly paternalist, as if epidemiologists were intellectually impaired or heart-weak. The main idea is easy to understand: To create causal relationships without knowing about the true nature of these causes, something that G.E.M. Anscombe claimed. If somebody is able to predict, intervene and manipulate the world she/he then does not need to know it ontologically. Then, language must be modified: Instead of causes, we should talk about causative meaning. They even used Wittgenstein's ideas to justify this domestication of causal controversies. Finally, there is a defense of probabilistic models of causation versus deterministic ones (from pragmatism to realism). As a consequence, the 2004 *Surgeon General's Report* used a new language (here semantics is the key):

"A. Evidence is *sufficient* to infer a causal relationship.
B. Evidence is suggestive but not sufficient to infer a causal relationship.
C. Evidence is *inadequate to infer* the presence or absence of a causal relationship (which encompasses evidence that is sparse, of poor quality, or conflicting).
D. Evidence is *suggestive of no causal* relationship."

After fierce debates on causality, the solution is to hide the debate and to adopt a pragmatic approach to statistics, concluding (Lipton and Ødegaard 2005: 9) that "for epidemiology, in particular, and science generally, the devil is in the details." Trying to find at least some sources of causative power, we could consider three levels (hierarchically placed), each one with specific and independent statistical analysis:

 i. Social epidemiology: social macrolevel (Berkman and Kawachi 2000).
 ii. Risk factor epidemiology: on behaviors and exposure, at the individual level.
iii. Molecular epidemiology: at the molecular microlevel.

The notion of "mechanism" as a partial and non-definitive approach to the reality pushed this field to a new view on causal models, between associationism and causal forces, but always through statistical assumptions. Causes are also substituted by "determining conditions" (Vineis 2003) or "active agents of change" (Kaufman and Poole 2000: 104). Renton (1994) calls this approach a "realist view on epidemiology and causation." This must now force us to abandon realist positions or the belief in solid facts that can be understood, and it only increases the complexity of the object under analysis at the same time that asks us to be open to several epistemological strategies (Wilkinson and Marmot 2003).

Finally, there is a last step of research on epidemiology: epidemiology in e-science environments (Thew 2009). New big data environments as well as computationally intensive methodologies represent a challenge to epidemiologists. Besides, the increasing implementation of e-science environment asks for the unification of language and, especially, of the tacit knowledge employed in reasoning about epidemiological data. New computational techniques will made possible to epidemiology to enter into a new level of performance, but not before making adjustments into their technical aspects.

As a conclusion to this chapter: If epidemiology was the contemporary cradle and test bed of new statistical techniques (regression models at the beginning or complex systems dynamic models in the last decades, Galea et al. 2010), this field went across a single cause model to a multicausal one, asking for more powerful techniques than those that frequentism could provide. Again, Bayesian statistics was the best option to deal with such complex analysis (Nguefack and Zucchini 2011). If it is true that frequentist tools were useful at the beginning of epidemiology, in the context of randomized trials and random-sample survey as the use of the methods spread from designed surveys and experiments to observational studies, why then did an increasing number of statisticians question the objectivity and realism of the hypothetical infinite sequences invoked by frequentist methods? (Rothman et al. 2008, Chap. 18). But in both approaches, the notion itself of "cause" has been silenced under the makeup of mechanistic and/or statistic approaches. Salmon-Dowe would be more interested on how mechanisms are based on physical facts, while Suppes, Lewis, or Price could be more focused into the idea of change (as difference-making) from a probabilistic, agent theory or counterfactual activity. Thus, the quantitative and the qualitative methods in epidemiology are related (Weed 2000).

Meanwhile, Bayesian methods have become common in advanced training and research in statistics (e.g., Leonard and Hsu 1999; Carlin and Louis 2000; Gelman et al. 2003; Efron 2005), even in the randomized trial literature for which frequentist methods were developed (e.g., Spiegelhalter et al. 1994, 2004). Following Broadbent (2013: 182), we could conclude that "the common theme through much of this discussion has been the importance of explanation and the relative de-emphasis of causation," but nothing further from the truth: Explanations are statistically flavored and they are founded in causal bonds between events of the world. What has changed radically has been the kind of tools that have been applied into epidemiological studies as well as the meaning of the notion of "cause." Causes are now open to include also complex social factors such as specific zone of residence, cultural habits or hobbies, among a long, rich, and sometimes discouraging variables list. Because we cannot forget that inductive reasoning, causes, predictions risks, or stable inferences are still live concepts into the field. Explanations are empty without causal and predictive power. Miracles and myths are powerful but fake explanations, and if we do not want to follow a fallacious ad honorem attitude toward epidemiologists, we must conclude that the range of the notion of causal relationship has been enriched by new meanings and relations. We move from close-controlled data environments to big data, fuzzy, complex, and open-uncontrolled ones. We are experiencing a second-level stochastication: The numerical control that we previously tried to establish between some variables and the rules that managed certain events have now been applied to the infinitesimal range of variables. We have a new level of statistic complexity that affects directly the variables weight before entering into how a cluster of variables affect certain problems. We have several webs inside a web, following a structure of a fractal statistics puzzle. It does not mean that we abandon or reject the possibility of causal claims, but that we are not so simple than before when we tried to find more complex causal patterns among huge amounts of data. It is a new step into the delicacies of epistemic knowledge of the world.

There is a corollary, at least for epidemiological researchers: Causes are related to behaviors, and the consequences of all events are interconnected following the butterfly effect, and numbers are ethically embedded. There is a strong connection between causes and consequences, and the same ethical project (or projects) becomes statistical (Vallverdú and Gustafsson 2009).

References

Andersen, H. (2007). History and philosophy of modern epidemiology based on a talk delivered at the &HPS conference, Pittsburgh, October 2007 from http://philsci-archive.pitt.edu/4159/1/Andersen_Modern_Epidemiology.pdf. Accessed March 7, 2014.

Bayes, T. (1763). An essay towards solving a problem in the doctrine of chances. *Philosophical Transactions of the Royal Society of London, 53*, 370–418.

Berkman, L. F., & Kawachi, I. (Eds.). (2000). *Social epidemiology*. NY: OUP.

Bertrand, J. L. F. (1889). *Calcul des probabilités*. Paris: Gauthier-Villars et fils.

Bhopal, R. S. (2002). *Concepts of epidemiology: An integrated introduction to the ideas, theories, principles and methods of epidemiology*. Great Britain: OUP.

Blackburn, H., & Labarthe, D. (2012). Stories from the evolution of guidelines for causal inference in epidemiologic associations: 1953–1965. *American Journal of Epidemiology, 176*(12), 1071–1077.

Broadbent, A. (2013). *Philosophy of the epidemiology*. London: Palgrave Macmillan.

Buck, C. (1975). Popper's philosophy for epidemiologists. *International Journal of Epidemiology, 4*(3), 159–168.

Carlin, B., & Louis, T. A. (2000). *Bayes and empirical-Bayes methods of data analysis*. NY: Chapman and Hall.

Cartwright, N. (1989). *Nature's capacities and their measurement*. Oxford: OUP.

Croce, P. (1996). The four postulates of Robert Koch. *Rivista di Biologia, 89*(2), 275–278.

Doll, R., & Hill, A. B. (1950). Smoking and carcinoma of the lung: Preliminary report 1950. *British Medical Journal*, 739–748.

Doll, R., & Hill, A. B. (1956). Lung cancer and other causes of death in relation to smoking. *British Medical Journal, 2*(5001), 1071–1081.

Efron, B. (2005). Bayesians, frequentists, and scientists. *Journal of American Statistical Association, 100*, 1–5.

Galea, S., Riddle, M., & Kaplan, G. A. (2010). Causal thinking and complex system approaches in epidemiology. *International Journal of Epidemiology, 39*, 97–106.

Gelman, A., Carlin, J. B., Stern, H. S., & Rubin, D. B. (2003). *Bayesian data analysis*. NY: Chapman and Hall/CRC.

Gillies, D. (2000). Varieties of propensity. *British Journal or Philosophy of Science, 51*(4), 807–835.

Greenland, S. (2006). Bayesian perspectives for epidemiological research I: Foundations and basic methods. *International Journal of Epidemiology, 35*, 765–775.

Greenland, S., Pearl, J., & Robins, J. M. (1999). Causal diagrams for epidemiologic research. *Epidemiology, 10*, 37–48.

Hill, A. B. (1965). The environment and disease: Association or causation? *Proceedings of the Royal Society of Medicine, 58*, 295–300.

Höfler, M. (2005). The Bradford Hill considerations on causality: A counterfactual perspective. *Emerging Themes in Epidemiology, 2*(11), 1–9.

Inglis, T. J. (2007). Principia aetiologica: Taking causality beyond Koch's postulates. *Journal of Medical Microbiology, 56*(11), 1419–1422.

Kaufman, J. S., & Poole, C. (2000). Looking back on 'causal thinking in the health sciences'. *Annual Review of Public Health, 21*, 101–119.

Krieger, N. (1994). Epidemiology and the web of causation: has anyone seen the spider? *Social Science and Medicine, 39*, 887–903.

Krieger, N. (2011). *Epidemiology and the people's health: Theory and context: Theory and context.* USA: OUP.

Leonard, T., & Hsu, J. S. J. (1999). *Bayesian methods.* Cambridge: CUP.

Lilienfeld, A. M. (1957). Epidemiological methods and inferences in studies of non-infectious diseases. *Public Health Review, 72*(1), 51–60.

Lipton, R., & Ødegaard, T. (2005). Causal thinking and casal language in epidemiology: It's in the details. *Epidemiologic Perspectives & Innovations, 2*, 1–9.

MacMahon, B., Pugh, T. F., & Ipsen, J. (1960). *Epidemiologic methods.* Boston: Little Brown and Co.

Marshall, G., et al. (1948). Streptomycin treatment of pulmonary tuberculosis: A medical research council investigation. *British Medical Journal, 2*(4582), 769–782.

McGrayne, S. B. (2011). *The theory that would not die: How Bayes' rule cracked the enigma code, hunted down Russian submarines, and emerged triumphant from two centuries of controversy.* USA: Yale University Press.

Morabia, A. (1991). On the origin of Hill's causal criteria. *Epidemiology, 2*(5), 367–369.

Nguefack-Tsague, G., & Zucchini, W. (2011). *Modeling hierarchical relationships in epidemiological studies: A Bayesian networks approach.* MPRA Paper No. 28232, at http://mpra.ub.uni-muenchen.de/28232/. Accessed March 9, 2014.

Olsen, J., Saracci, R., & Trichopoulos, D. (2010). *Teaching epidemiology: A guide for teachers in epidemiology, public health and clinical medicine.* Bangalore: OUP.

Phillips, C. V., & Goodman, K. J. (2004). The missed lessons of Sir Austin Bradford Hill. *Epidemiologic Perspectives & Innovations, 1*(3), 1–5.

Renton, A. (1994). Epidemiology and causation: A realist view. *Journal of Epidemiology and Community Health, 48*, 79–85.

Rothman, K. J. (1976). Causes. *American Journal of Epidemiology, 104*, 587–592.

Rothman, K. J., Greenland, S., & Lash, T. L. (2008). *Modern epidemiology.* Philadelphia, PA: Lippincott, Williams & Wilkins.

Spiegelhalter, D. J., Abrams, K. R., & Myles, J. P. (2004). *Bayesian approaches to clinical trials and health-care evaluation.* NY: Wiley.

Spiegelhalter, D. J., Freedman, L. S., & Parmar, M. K. B. (1994). Bayesian approaches to randomized trials (with discussion). *Journal of the Royal Statistical Society Series A, 156*, 357–416.

Suárez, M. (Ed.). (2011). *Probabilities, causes and propensities in physics.* Dordrecht: Springer.

Suppes, P. (1970). *A probabilistic theory of causality.* Helsinki: North-Holland.

Susser, M. (1973). *Causal thinking in the health sciences: Concepts and strategies of epidemiology.* New York: Oxford University Press.

Susser, M. (1991) What is a cause and how do we know one? A grammar for pragmatic epidemiology. *American Journal of Public Health, 133*(7), 635–648.

Susser, M. (2001). Glossary: Causality in public health science. *Journal of Epidemiology and Community Health, 55*, 376–378.

Susser, E. (2004). Eco-epidemiology: Thinking outside the black box. *Epidemiology, 15*, 519–520.

Susser, M., & Susser, E. (1996a). Choosing a future for epidemiology I: Eras and paradigms. *American Journal of Public Health, 86*, 668–673.

Susser, M., & Susser, E. (1996b). Choosing a future for epidemiology II: From black box to Chinese boxes and eco-epidemiology. *American Journal of Public Health, 86*, 674–677.

Thew, S., et al. (2009). Requirements engineering for e-science: Experiences in epidemiology. *IEEE Software, 26*(1), 80–87.

Vallverdú, J., & Gustafsson, C. (2009). Synthetic life: Etho-bricks for a new biology. In P. Fu & S. Panke (Eds.), *Systems biology and synthetic biology* (pp. 205–220). New York: Wiley.

Vandenbroucke, J. P. (2002). The history of confounding. *Sozial- und Präventivmedizin, 47*, 216–224.

Vineis, P. (2002). History of bias. *Sozial- und Präventivmedizin, 47*, 156–161.

Vineis, P. (2003). Causality in epidemiology. *Soz-Präventivmedizin, 48*, 80–87.

Weed, D. L. (1998). Beyond black box epidemiology. *American Journal of Public Health, 88*(1), 12.

Weed, D. L. (2000). Interpreting epidemiological evidence: How meta-analysis and causal inference methods are related. *International Journal of Epidemiology, 29*, 387–390.

Wilkinson, R., & Marmot, M. (Eds.). (2003). *Social determinants of health: The solid facts*. Copenhagen: WHO. Accessible online http://www.euro.who.int/__data/assets/pdf_file/0005/98438/e81384.pdf

Yerushalmy, J., & Palmer, C. E. (1959). On the methodology of investigations of etiologic factors in chronic diseases. *Journal of Chronic Diseases, 10*, 27–40.

Chapter 7
Natural Versus Artificial Minds and the Supercomputing Era

Abstract Computer sciences have completely changed the way scientific and social research is performed nowadays. This chapter analyzes the role of Bayesianism and frequentism into the emergence of e-science, artificial intelligence, and robotics, the generation of expert systems, and the overwhelming problem of how to analyze Big Data, a process called "data mining." This review of the main systems and ideas will show us how Bayesianism is acquiring a determinant position among worldwide users of statistical tools.

Keywords AI · E-science · Robotics · Data mining · Expert system · Supercomputing · Big data · Cognitive systems

In May of 1623, Galileo Galilei published his famous book *Il Saggiatore, nel quale con bilancia squisita e giusta si ponderano le cose contenute nella Libra*. In Sect. 6 can be found one of his most famous quotes (see footnote 1):

> La filosofia è scritta in questo grandissimo libro che continuamente ci sta aperto innanzi a gli occhi (io dico l'universo), ma non si può intendere se prima non s'impara a intender la lingua, e conoscer i caratteri, ne' quali è scritto. Egli è scritto in lingua matematica, e i caratteri son triangoli, cerchi, ed altre figure geometriche, senza i quali mezi è impossibile a intenderne umanamente parola; senza questi è un aggirarsi vanamente per un oscuro laberinto.[1]

Mathematics and science since then share a common flavor. Once sciences evolved and their complexity increased, statistical methodologies were necessary to solve or treat several problems. After this, the number of data became so high that we needed to use computational power to process this great amount of information. Therefore, sciences, mathematics, statistics, and computers are deeply related to each other in our time. Bigger data sets asked for data-mining techniques and more complex methods were incorporated into statistical practices.

---

[1]Here is the English translation, by Drake and O'Malley (1960): "Philosophy is written in this grand book, the universe, which stands continually open to our gaze. But the book cannot be understood unless one first learns to comprehend the language and read the letters in which it is composed. It is written in the language of mathematics, and its characters are triangles, circles, and other geometric figures without which it is humanly impossible to understand a single word of it; without these, one wanders about in a dark labyrinth."

© The Author(s) 2016
J. Vallverdú, *Bayesians Versus Frequentists*,
SpringerBriefs in Statistics, DOI 10.1007/978-3-662-48638-2_7

This is our last chapter of compared analysis of Bayesian and frequentist statistical methods, and we will focus our research into a very important domain: computer sciences. Computer facilities are the backbone of contemporary scientific research. Not only in the cases of e-science or e-humanities, but also in continuing classic research run together with computing power and techniques (Vallverdú 2009; Casacuberta and Vallverdú 2014). At the same time, not only has science been increasingly computerized but computer sciences also have played a very important role in statistical data analysis.

7.1 Artificial Intelligence and Statistics

AI is a multidisciplinary research field with broad and ambitious interests regarding artificial cognitive systems. The necessity of working with intensive and sometimes pervasive numerical data as well as artificial cognitive systems requires from statistical techniques ability to deal with those problems. Machine supervised or unsupervised learning, knowledge discovery, data mining,[2] simulations, analytics, or predictive modeling have been a serious challenge for AI experts. Despite some frequentist implemented methodologies, we can affirm that the domination of Bayesian models into AI is overwhelmingly superior. A number of recent ACM Turing Award winners are close to Bayesian ideas. Among them, we would like to mention Judea Pearl, who championed the probabilistic approach in AI, developer of the Bayesian Networks[3] (Pearl 1999) as well as the creator of a theory of causal and counterfactual inference based on structural models (Pearl 2009).[4] Again, we

---

[2]Data mining is a crucial discipline in the twenty-first century, but must be run under solid techniques. Austin et al. (2006) have shown how intensive data mining can obtain invalid results, like the false relationships they obtained associating causally astrological signs and health. The nontrivial extraction of implicit, previously unknown, and potentially useful information from data is not free from error. In their case despite of adding a validation cohort study, the false association remained.

[3]Following Gillies (1998, as well as personal communications), it is true that Bayesian networks can be interpreted entirely within the framework of classical (frequentist) statistics. This is done by (a) interpreting the probabilities objectively as frequencies or propensities and (b) testing the assumptions involved by statistical tests. It matches perfectly with spirit of Pearl, who is not interested in the purity of one school or another but on the uses and design of statistics. As happens with mathematical formulation and computer language programming, the same idea can be perfectly translated from one domain to another, if the symbolic system is powerful enough, as happens today with Bayesian and frequentist paradigms.

[4]Curiously, he admits that "the debate on "who is a Bayesian" is empty, and even those who pride themselves on being "Bayesian" do not really understand what drives them, and what justifies what they are doing," quoted from personal e-mail conversation (April 6, 2014). He even defines himself as "Half-Bayesian" (Pearl 2001: 19), considering probabilities as subjective but causes as objective. Again, the ontological constraint of the human (and human-externally supported or extended) observer position forces us to distinguish between our ideas and "the world itself," although we only can deal with ideas.

found the connection between pioneering into statistical nature of events, statistical methodologies, and causal thinking.

Computers and statistics have a close relationship, because the former allowed new ways to perform calculations and to analyze data. Let me show you a related example: in 1959, Robert Ledley wrote two influential articles in *Science*: "Reasoning Foundations of Medical Diagnosis" (with Lee B. Lusted) and "Digital Electronic Computers in Biomedical Science"[5] which inspired the next generation of physicians to use computational facilities in their research. Both articles encouraged biomedical researchers and physicians to adopt computer technology.[6] Just two years later, Homer Warner wrote the influential paper to explain the creation of the first computerized program for diagnosing disease. He used Bayesian analysis (Warner et al. 1961). Diagnosis of liver disease and congenital heart disease were among the first subjects for which computer-based Bayesian models were constructed. As soon as Bayesian-network technology became available at the end of the 1980s, biomedical researchers started developing Bayesian networks, usually using expert knowledge as a foundation. Examples of early Bayesian-network systems include Pathfinder, a system aimed at supporting pathologists in the diagnosis of white-blood-cell tumors, and MUNIN, a system meant to assist neurologists in the interpretation of electromyograms (Lucas 2004: 221).

The implementation of computational power into statistics domains has also allowed the emergence of new techniques that have been used by AI experts as well. For example, bootstrap is a frequentist machine that produces Neyman-like confidence intervals (Efron 2012: 140)[7]; on the other side, Bayesians methods have been also automated, for example with the *Gibbs sampling* (a Markov Chain[8]

---

[5]He also created the first whole-body computer tomography machine. See Sittig et al. (2006), for a nice and direct contact with this creation process.

[6]Very often, it has been affirmed that the higher technical complexity of Bayesian calculus before the computers era was one of the first practicals reasons against its main implementation. I can agree with that view, but not always simplicity or parsimony is the best approach to scientific truth. For a complete analysis of this idea, I suggest the reading of Foster and Sober (1994), Schurz (2015) prefers to look at statistical "unification power" at the theoretical level instead of simplicity.

[7]Joseph Felsenstein started in 1985 adapting bootstrap method of statistics to phylogenies, and this provided a way to know which aspects of the evolutionary tree were well-supported or not. Something that started as a personal hobby, an eccentricity, became an excellent and recognized technique some years later. (According to his own words, expressed at *JSPS Quarterly*, 47: 2).

[8]Andrei Andreevich Markov (1856–1922, father...his son was called exactly like him and was also a mathematician) was a Russian mathematician with multiple interests. According to Hayes (2013), Markov founded a new branch of probability theory by applying mathematics to poetry. Delving into the text of Alexander Pushkin's novel in verse *Eugene Onegin*, Markov spent hours sifting through patterns of vowels and consonants. On January 23, 1913, he summarized his findings in an address to the Imperial Academy of Sciences in St. Petersburg. His analysis did not alter the understanding or appreciation of Pushkin's poem, but the technique he developed—now known as a Markov chain—extended the theory of probability in a new direction. Markov's methodology went beyond coin flipping and dice-rolling situations (where each event is independent of all others) to chains of linked events (where what happens next depends on the current

random walk procedure). In 1989, Smith and Gelfand wrote a crucial paper in which they demonstrated how Markov Chain Monte Carlo Methods could be applied to almost any statistical problem, by replacing integration by MCMC. This led to a new world of possibilities for scientists and started the worldwide spreading of Bayesian methods. Consequently, specific software was created to deal with all these changes and BUGS (for Bayesian Statistics Using Giggs Sampling) is one of the most reputed. JASP is another powerful and open-source statistical software.[9]

Bayesian ideas are implemented in several computer environments such as spamming filtering, Web site searching, automated translations, data mining, and even in parts of the Microsoft software or forms part of new ideas on emotional AI when Bayesian methods are implemented into uncertainty management in emotion-based reasoning (Chakraborty and Konar 2009). Anyhow, Bayesian ideas are of great influence in several AI domains such as vision, natural language parsing, concept learning, or categorization.

7.2   Supercomputing, Big Science and Big Data

The emergence of the Big Data context, boosted by intensive computational facilities, has reached the practices of all scientific and humanistic disciplines. Supercomputers, distributed computing resources or even small grids of computational stations[10] are starting to deal with huge amounts of data (according to the research size and topic) that need to be collected, checked, stored, and analyzed. The statistics of Big Data is a new challenge for contemporary sciences and

---

(Footnote 8 continued)

state of the system). A Markov chain is a mathematical system under a random process that undergoes transitions from one state to another, with the property that the next state depends only on the current state (memoryless). Some decades later, and during the intense researches on atomic bombs at Los Alamos, the basic idea of Markov led to the birth of the Monte Carlo method, in 1949, when an article entitled "The Monte Carlo method" by Metropolis and Ulam appeared. The American mathematicians John von Neumann and Stanislav Ulam are considered its main originators (Sobol 1994). The resultant algorithm was also called Markov Chain Monte Carlo Method (MCMC) and has been widely and intensively used in very diverse disciplines, especially in physics (E. Fermi or M.G. Mayer) and chemistry. For a deep analysis of MCMC creation, see Robert and Casella (2011). RAND was one of the first believers in MCMC, and from this research the Hastings-Metropolis algorithm emerged later, to work with huge problems involving thousands of hypothesis and of parallel inference programs (Bertsch 2011: 223).

[9]https://jasp-stats.org/, accessed on May 24, 2015.

[10]At the beginning of the twentieth century, automated stations were created to control information exchange infrastructures, specially telephonic ones. Edward C. Molina worked for Bell company and applied Bayesian ideas to make possible the relay translator, saving Bell survival in a changing era.

industries.[11] In 2011, an IDC report defined Big Data as "Big Data technologies describe a new generation of technologies and architectures, designed to economically extract value from very large volumes of a wide variety of data, by enabling the high-velocity capture, discovery, and/or analysis."[12] With this definition, characteristics of Big Data may be summarized as four Vs related to information, i.e., Volume (great volume), Variety (various modalities), Velocity (rapid generation), and Value (huge value but very low density).[13] DARPA's *Big Mechanism* program[14] allows researchers to find new information contained in various scientific reports and papers published around the world and then absorb that information into their ongoing work. As they declare[15]: "The Big Mechanism program aims to develop technology to read research abstracts and papers to extract pieces of causal mechanisms, assemble these pieces into more complete causal models, and reason over these models to produce explanations. The domain of the program is cancer biology with an emphasis on signaling pathways. Although the domain of the Big Mechanism program is cancer biology, the overarching goal of the program is to develop technologies for a new kind of science in which research is integrated more or less immediately—automatically or semi-automatically—into causal, explanatory models of unprecedented completeness and consistency. Cancer pathways are just one example of causal, explanatory models."

Among the list of existing supercomputers as well of projects involved in Big Data analysis, Bayesianism is a leading option. Thanks to new classes of Monte Carlo inference procedures for statistical inference it is possible to scale billions of data items. Bayesian learning with Big Data is now the best option,[16] as well as for a long list of computational activities such as artificial intelligence, machine learning, analytics, or deep learning[17]. For example, as Chai et al. (2013) explain Bayesian inference is one of the most important methods for estimating phylogenetic trees in bioinformatics (See also Reumann 2012). We could show dozens of examples following the same mood and content: the overall superiority of Bayesian techniques into supercomputing and Big Data environments. Could this vast

---

[11]See Chen et al. (2014). As they point "For example, Google processes data of hundreds of Petabyte (PB), Facebook generates lots of data of over 10 PB per month, Baidu, a Chinese company, processes data of tens of PB, and Taobao, a subsidiary of Alibaba, generates data of tens of Terabyte (TB) for online trading per day."

[12]See http://www.emc.com/collateral/analyst-reports/idc-extracting-value-from-chaos-ar.pdf, accessed on March 14, 2014.

[13]Ibid.

[14]See https://www.fbo.gov/index?s=opportunity&mode=form&id=fe17239c6586a4d6521a09ad3-a7aa5b7&tab=core&_cview=0, accessed on April 27, 2014.

[15]http://www.darpa.mil/program/big-mechanism accessed on September 1, 2015.

[16]See Yuan (Alan) Qi, "Bayesian learning with Big Data," from: https://www.cs.purdue.edu/homes/alanqi/papers/Qi-UCL-July-2-2012.pdf. Accessed on April 8, 2014.

[17]Deep learning is the next step in the evolution of AI and contemporary data sciences, and its spirit is 100 % Bayesian. This research group provides good references on this topic: http://research.ics.aalto.fi/bayes/, accessed June 1, 2015.

amount of data allow a frequentist interpretation? My answer is still "no," because the key point is not the size of raw data but the epistemological approach. And Bayesian allows a more dynamic and intuitive approach to new data.

While classic symbolic AI approaches like CYC[18] were not interested on Bayesian or human-like reasoning, some pioneering expert systems like PROSPECTOR used Bayesian reasoning. At the same time, new AI trends have been shifting toward Bayesian approaches (Spiegelhalter 1993). Created for company publicity interests as well as for research purposes, IBM's Watson supercomputer defeated all time Jeopardy champions Ken Jennings and Brad Rutter in February 2011. Although Watson ran on IBM's own software, called DeepQA, this software used a variety of weighting schemes including Bayesian techniques to assess the accuracy of its answers (Ferrucci et al. 2010).[19]

Again on Big Data, contemporary e-science and Big Commerce strategies as well as Social Networks' environments are increasingly working with huge amounts of data. Therefore, Big Data Analytics has become a fundamental topic of research. Using machine learning tools such as Bayesian nonparametrics, an area in machine learning, in which models grow in size and complexity as data accrue, can help to deal with this. But their algorithms for posterior inference generally show poor scalability and are extremely slow when analyzing massive amounts of data. People like John Paisley are working on a general and efficient variational inference strategy for learning based on stochastic optimization, and he shows that with this combination of modeling and inference approach, we are able to learn high-quality models using millions of documents.

References

Austin, P. C., et al. (2006). Testing multiple statistical hypotheses resulted in spurius associations: A study of astrological signs and health. *Journal of Clinical Epidemiology, 59*, 964–969.

Baum, E. B., & Smith, W. D. (1997). A Bayesian approach to relevance in game-playing. *Artificial Intelligence, 97*, 195–242.

Campbell, M. S., Hoane, A. J., & Hsu, F. (1999). *Search control methods in deep blue*. AAAI Technical Report SS-99-07, 19–23.

Casacuberta, D., & Vallverdú, J. (2014). E-science and the data deluge. *Philosophical Psychology, 27*(1), 126–140.

Chai, J., et al. (2013). Resource-efficient utilization of CPU/GPU-based heterogeneous supercomputers for Bayesian phylogenetic inference. *The Journal of Supercomputing, 66*(1), 363–380.

Chakraborty, A., & Konar, A. (2009). *Emotional intelligence*. Berlin: Springer.

---

[18]See http://www.cyc.com/documentation/overview-cyc-inferencing/, accessed on September 1, 2015.

[19]14 years before IBM's Deep Blue supercomputer defeated human chess world champion Gary Kasparov in a six-game match, using some forms of alpha–beta pruning algorithms (Campbell et al. 1999), not directly related to Bayesian thinking but that yielded to Bayesian approaches to games (Baum and Smith 1997). And Bayesian thinkers like J. Pearl worked on them: Pearl (1982).

Chen, M., Mao, S., & Liu, Y. (2014). Big data: A survey. *Mobile Network Applications*. doi: 10. 1007/s11036-013-0489-0, from http://mmlab.snu.ac.kr/~mchen/min_paper/BigDataSurvey-2014.pdf. Accessed on March 14, 2014.

Drake, S., & O'Malley, C. D. (1960). *The controversy on the comets of 1618* (p. 263). USA: University of Pennsylvania Press.

Efron, B. (2012). A 250-year argument: Belief, behavior, and the bootstrap. *Bulletin American Mathematical Society, 50*, 129–146.

Ferrucci, D., et al. (2010). Building watson: An overview of the DeepQA project. *AI Magazine, 3*(4), 59–79.

Foster, M., & Sober, E. (1994). Hot to tell when simpler, more unified or less ad hoc theories provide more accurate predictions. *British Journal of the Philosophy of Science, 45*, 1–35.

Gillies, D. (1998). Debates on Bayesianism and the theory of Bayesian networks. *Theoria, 64*(1), 1–22.

Hayes, M. (2013). First links in the Markov Chain. *American Scientist, 101*, 92–97.

Lucas, P. J. F. (2004). Restricted Bayesian network structure learning. In J. A. Gómez, S. Moral, & A. Salmeron (Eds.), *Advances in Bayesian networks, studies in fuzziness and soft computing* (Vol. 146, pp. 217–232). Berlin: Springer.

McGrayne, S. B. (2011). *The theory that would not die: How Bayes' rule cracked the enigma code, hunted down Russian submarines, and emerged triumphant from two centuries of controversy.* USA: Yale University Press.

Pearl, J. (1982). The solution for the branching factor of the alpha-beta pruning algorithm and its optimality. *Communications of the ACM, 25*(8), 559–564.

Pearl, J. (1999). *Causality: Models, reasoning, and inference.* UK: CUP.

Pearl, J. (2001). Bayesianism and causality, or, why I am only a half-Bayesian. In D. Corfield & J. Williamson (Eds.), *Foundations of Bayesianism* (pp. 12–36). The Netherlands: Kluwer.

Pearl, J. (2009). Causal inference in statistics: An overview. *Statistics Surveys, 3*, 96–146.

Reumann, M. (2012). Supercomputing enabling exhaustive statistical analysis of genome wide association study data: Preliminary results. In *Conference Proceedings of the IEEE Engineering Medicine and Biology Society* (pp. 1258–1261).

Robert, C., & Casella, G. (2011). A short history of Markov Chain Monte Carlo: Subjective recollections from incomplete data. *Statistical Science, 26*(1), 102–115.

Schurz, G. (2015). Causality and unification: How causality unifies statistical regularities. *Theoria, 30*(1), 73–95.

Sittig, et al. (2006). The story behind the development of the first whole-body computerized tomography scanner as told by Robert S. Ledley. *Journal of the American Medical Informatics Association, 13*(5), 465–469.

Sobol, I. M. (1994). *A primer for the Monte Carlo method.* Raton (FL): CRC Press.

Spiegelhalter, D. J., et al. (1993). Bayesian analysis in expert systems. *Statistical Science, 8*(3), 203–353.

Vallverdú, J. (2009). Computational epistemology and e-science: A new way of thinking. *Minds and Machines, 19*(4), 557–567.

Warner, H. R., Toronto, A. F., Veasey, L. G., & Stephenson, R. (1961). A mathematical approach to medical diagnosis: Application to congenital heart disease. *JAMA: The Journal of the American Medical Association, 177*(3), 177–183.

Chapter 8
And the Winner Is...

Abstract The concluding chapter of this book makes a revision about the way by which scientists choose between ideas or techniques and how it is important for statistics. The notion of epistemological opportunism and the consideration of sciences as problem-solving activities make it possible to understand the fierce statistical debates as well as show us a way to find a winner to the contest: Bayesianism. At the same time, the possibility of the notion of cause is defended under a mixed approach to statistical tools.

Keywords Problem solving · Epistemology · Winner · Causality · Pearl · Epistemic opportunism

After our long journey across the main schools, ideas and protagonists of the statistical thinking, we must ask ourselves if there is a winner, whether somebody has the better solution to the epistemological problems humans are faced with in our time. To offer you an answer I will take a metaphor from biological tradition: there is not a clear winner, just the most adapted and fittest to the challenges of contemporary sciences, and it is... *Bayesianism.*[1]

As Cousins (1995: 198) notes, a Nobel Laureate in condensed matter theory decreed matter-of-fact that Bayesian statistics "are the correct way to do inductive reasoning from necessarily imperfect experimental data," and is useful in order to solve several scientific problems, something that has propitiated the idea of "therapeutic Bayesianism" (Horwich 2005). My humble vision on this topic is clear: not only Bayesians have showed to be more powerful, versatile and decisive

---

[1] I recommend also to read the 10 specific advantages of Bayesianism listed by Wagenmakers et al. (2008). They are very interesting and, in some cases, even funny. It is the kind of sardonic humor with a little of lack of respect toward the contrary that we usually find at the frequentist side. In any case, Bayes' Theorem can even be studied with Lego pieces: https://www.countbayesie.com/blog/2015/2/18/bayes-theorem-with-lego, accessed in May 24, 2015.

© The Author(s) 2016
J. Vallverdú, *Bayesians Versus Frequentists*,
SpringerBriefs in Statistics, DOI 10.1007/978-3-662-48638-2_8

in most complex scientific environments[2] but also have demonstrated an open attitude to include frequentist ideas and methods into their models, while frequentist experts are usually horrified with Bayesian techniques. The former is the basic way of evolution in science.

Anyhow, at the same time we have seen that blurred lines exist between both views (a broad range of gray tonalities between clear black and white) as well as their practical implementations often justify combined methodologies. Scientists and methodologies have a curious relationship, and according to Einstein (1949: 683–684),[3] they are unscrupulous opportunist.[4] As Platonist or Pythagorean, the scientist furnishes a logical representation of relations that can or not come from direct sensory experiences.

If we forget the ontological debate on the foundation of both main schools and accept that numbers do not exist beyond our minds and models, we are clearly faced with an undetermination situation (Duhem–Quine thesis): several theories can explain by different methods and ways the same phenomenon. If we consider sciences as "problem-solving" disciplines (Funtowicz and Ravetz 1994), it is enough with being able to deal with certain areas of reality and to receive good feedback. I am not postulating toward a new saving the phenomena attitude in science, but it is necessary to admit that statistical tools are instruments to create meaning about the world…they are not the meaning, just the messenger…and we

---

[2]According to McGrayne (2011), the whole twentieth century would be completely different without Bayesianism: Alan Turing used it to break the Enigma Code (with a Bayesian system he nicknamed Banburismus; there was also a Bayesian method called Turingery or Turingismus), Andre Kolmogorov (in Russia, also in military-related research), and Claude Shannon (in the United States) rethought Bayes for wartime decision making, helped to find lost bombs and submarines, is in Microsoft, Google, Wall Street… for her it's the panacea. It is true not only from a practical or applied point of view but also as a successful metaphor about brain performing.

[3]About simplicity and Bayesianism, I must recommend the reading of the dissertation of Scoto (2003).

[4]Following Levine and Perlovsky (2010: 1), we must point out that Leven (1987) and Levine (1998, Chap. 7) posited three major problem solving styles, each named after a different mathematician whose major work illustrated the essence of that style. Their three types are "Dantzig" or direct solvers who try simply to achieve an available solution by a repeatable method; "Bayesian" solvers who play the percentages and try to maximize a measurable criterion; and "Godelians" who use both intuition and reason to arrive at innovative solutions. The Bayesian solver type is the one idealized in normative decision theory and is the best suited to problems of the ratio-bias or base-rate type. Yet Godelian solvers are better at problems that are much more open-ended, such as designing the best possible office environment; hence, they are often valued in group brainstorming situations. Even in quantifiable domains, the Godelian tendency frequently leads great thinkers to find solutions that highly competent Bayesians have overlooked. For example, Albert Einstein was led to his relativity theory by cognitive dissonance between some new (relatively minor) results on light and radiation and the Newtonian paradigms for physics (Cline 1965). Before him even other great physicists such as Planck had largely glossed over the discrepant data. Poincaré (1914) described the process of discovery of mathematical proofs as involving alternating periods of logical deduction and intuitive insight. No truth in a single repository of ideas, nor evolution from a closed set of conceptual rules or data.

do not need to kill the messenger, or none of them! This does not justify relativism,[5] but a moderate realism embedded in operationalism. This is the kind of illustrated opportunism to which Einstein pointed to. There is even a deeper problem: statistics manipulation even if its nature is as an over-confident discipline (Funtowicz and Raven 1990). If data and models are not good, and at the same time there is a general pressure toward the use of statistical tools, then you obtain a GIGO pseudo-science (Garbage In, Garbage Out). The idea of Funtowicz and Ravetz is to affirm that no single system will be able to prevent abuse and corruption of statistics, especially when we are dealing with uncertainty.[6] Just collecting numbers is not enough, but at the same time the pursuit of a single methodology or technique to obtain pure and definitive results is a chimera. Perhaps not banished, but uncertainty must be managed with a multiplicity of conceptual tools, completely immersed into a post-normal science paradigm.

We can explain this looking at classic human epistemology: we do not discuss about which are the best senses to capture the world because we are naturally equipped with specific sensorial tools to capture sections of the world. With the extra aim of external instruments, we are able to check, enrich, and analyze that reality that is beyond our basic senses. At the same time, a holistic capture of the reality is not possible, because there are several levels of activity and order, each one of them operated by different mechanistic processes: quantum events are not similar to those of macrophysics, and systems biology, for example, they deal with specific living processes: different layers, different mechanisms, different methodologies. Once we have all the information, we can try to see the whole forest and not only a huge amount of trees.[7] In this moment, Bayesian statistics is widely accepted and provides powerful and successful methodologies for sciences (and

---

[5]As made Paul Feyerabend with his lemma *Anything Goes*. His works are basic for any reader with interests on scientific epistemology, with titles like *Against Method: Outline of an Anarchistic Theory of Knowledge* (1975) or *Farewell to Reason* (1987). As Alexander Pope (1688–1744) said, "False eloquence, like the prismatic glass,/ Its gaudy colors spreads on every place"; according to this spirit, Feyereband thought that scientific facts were social constructions and that observations were the result of interventions. Then to know is not a possible filtering of truth, but a direct action over the reality. As a good epistemological anarchist, Feyerabend considered that there was not a unified method for all sciences, and instead of it, scientists behave like epistemic opportunists.

[6]I need the reader to remember the famous categorization of Wynne (1992), where he defined four different levels: (i) risk for known odds, (ii) uncertainty for unknown odds, (iii) ignorance for lack of understanding about what we really know, and (iv) indeterminacy, for causal chains or networks open. The uncertainty can provide from incomplete or imperfect observations, from incomplete conceptual frameworks, from inaccurate prescriptions of known processes (by poor parametrisations, etc.), by intrinsic chaos or by intrinsic lack of predictability.

[7]I use again a quote from A. Einstein: "I fully agree with you about the significance and educational value of methodology as well as history and philosophy of science. So many people today— and even professional *scientists—seem to me like somebody who has seen thousands of trees but has never seen a forest*. A knowledge of the historic and philosophical background gives that kind of independence from prejudices of his generation from which most scientists are suffering. This independence created by philosophical insight is—in my opinion—the mark of distinction between a mere artisan or specialist and a real seeker after truth." A. Einstein to R.A. Thornton,

economics), but frequentist contributions illuminate other areas of the reality, creating very often complementary views about possible truths. Immanuel Kant once used the metaphor of the truth as island surrounded by fog or icebergs as illusions, which needed to be illuminated by reason.[8] The world is much more complex, because there are many islands, an archipelago that we need from all ways and methods to discover. The voyage of reason can be done upon the shoulders of statisticians, but only with those who have an open-minded attitude. We need to be methodological pluralists. Our instruments cannot become dogmas, as Lovric (2011) included into his recent *International Encyclopedia of Statistical Science*.

There is also a necessary conceptual remark: the gaps between theoreticians–philosophers and the practitioners of statistical methodologies are really big. Thinkers detail the ontological and epistemological claims they consider necessary, while statisticians try to justify better data processing, avoiding at a certain point the conceptual debate, despite from time to time burning heavily in strong quarrels. Hilariously, for the mathematician John Warren Tukey the collective noun for a group of statisticians was a *quarrel*. The explained thesis of the operationalism can overcome this problem, although they do not solve it definitively. If it works, then it is fine, don't worry. But at the same time, I need to express my ideas about the philosophical debates on probability: as far as I can see, the several theories, mental experiments, countertheories, and overdetailed analysis on causality, retrocausality, multicausality, propensions, priors or long-run sequences, are all of them fictional modelizations about reality: sciences deal with a vast range of phenomena that must be managed. For those purposes, a broad number of techniques exist that run under basic methodological agreements on accuracy, quantification, honesty, and coherence. At a certain level, anything goes if follows these basic principles. We have seen that several approaches, and even mixed- or cross-techniques, work correctly

---

(Footnote 7 continued)

unpublished letter dated December 7, 1944 (EA 6-754), Einstein Archive, Hebrew University, Jerusalem cited by Howard (2005).

[8]*The Critique of Pure Reason*, Chap. 3. Of the Ground of the Division of all Objects into Phenomena and Noumena: "We have now not only traversed the region of the pure understanding and carefully surveyed every part of it, but we have also measured it, and assigned to everything therein its proper place. But this land is an island, and enclosed by nature herself within unchangeable limits. It is the land of truth (an attractive word), surrounded by a wide and stormy ocean, the region of illusion, where many a fog-bank, many an iceberg, seems to the mariner, on his voyage of discovery, a new country, and, while constantly deluding him with vain hopes, engages him in dangerous adventures, from which he never can desist, and which yet he never can bring to a termination. But before venturing upon this sea, in order to explore it in its whole extent, and to arrive at a certainty whether anything is to be discovered there, it will not be without advantage if we cast our eyes upon the chart of the land that we are about to leave, and to ask ourselves, firstly, whether we cannot rest perfectly contented with what it contains, or whether we must not of necessity be contented with it, if we can find nowhere else a solid foundation to build upon; and, secondly, by what title we possess this land itself, and how we hold it secure against all hostile claims? Although, in the course of our analytic, we have already given sufficient answers to these questions, yet a summary recapitulation of these solutions may be useful in strengthening our conviction, by uniting in one point the momenta of the arguments."

for specific purposes. No single method can provide a solution. I know that parsimony is necessary, and I agree with this principle. But atomization of statistical procedures in small and fragmented pieces broke the whole: nature is relational and works at different organizational levels. Following the ideas of Robert Rosen, we need to repeat that "the whole is more than the sum of parts." The diverse statistical methods offer good ways to solve problems and to interact better with the world: frequentist or Bayesian flavors are not black or white, but justify a wide number of numerical techniques to explain the world, some of them complementary. Like Zeno, Achilles, and the tortoise, we can get trapped by ideas that make us see only the trees and to forget the forest.

But I would like to close this book with one question and some final remarks: *Is there still a space for causality in statistics?* We have seen that the fights between statisticians are at the end not only a question of honor but also the result of the true belief into the existence of a unified methodology able to explain the real nature of the world. My answer to the last question in this book is affirmative, and I will explain to you why through the ideas of Pearl (2000, 2009). Most studies in the health, social, and behavioral sciences are not motivated by associational but causal elements in nature. They rely on the mechanisms that support events, not in the statistical distributions that govern the data. Besides, the standard statistical analysis (typified by regression, estimation, and hypothesis testing techniques) is working with static data, while causal analysis work with mechanisms that produce data under changing conditions. There is a deep ontological distance between associational and causal concepts[9]. The first ones provide from any relationship that can be defined in terms of a joint distribution of observed variables, if the sample is large enough, while the second ones cannot be defined from the distribution alone, but ask for identified relationships that remain invariant when external conditions change,[10] i.e., the causes, and can be verified by experimental control. Without taking this into account is not possible, for example, to see that confounding bias in frequentist statistics cannot be detected or corrected by statistical methods alone: some judgmental assumptions regarding causal relationships must be done in the problem before the adjustments can correct it. In order to obtain knowledge about

---

[9]As Pearl (2001: 28) notes "the slogan 'correlation does not imply causation' can be translated into a useful principle: one cannot substantiate causal claims from associations alone, even at the population level—behind every causal conclusions there must lie some causal assumption that is not testable in observational studies." Nancy Cartwright expressed it positively as "no causes in, no causes out," something that we can redefine more rudely as "garbage in, garbage out." Cartwright makes it possible to talk about statistical causality without implying determinism. As an example of very funny spurious correlation between margarine consumption and rate of divorces, see: http://www.bbc.com/news/magazine-27537142, and more examples at http://www.tylervigen. com/ accessed on July, 18, 2014. Anyhow, recent data are suggesting that at least at quantum level of analysis, quantum correlation can imply causation, see Ried et al. (2015).

[10]Pearl (2001: 36) discusses the insufficient approach of Suppes (1970), where the philosopher tried to justify that the calculus of probabilities endowed with a time dynamic would be sufficient for causation. There are counterexamples for the Suppes arguments, provided by Otte in 1981 about light perceptions and chains of delayed switches.

these causes, they could use existing conceptual and algorithmic tools but they are far from their educational skills. Four areas of maximum interest are not under the necessary study and application: (1) counterfactual analysis (we are forced to admit extra-probabilistic primitives), (2) nonparametric structural equations, (3) graphical models, and (4) symbiosis between counterfactual and graphical models. According to Pearl (2009), the transition from statistical to causal analysis is not yet reached because of two barriers: (a) coping with untested assumptions and (b) the lack of a suitable mathematical notation for these purposes.[11] Although statisticians do not think on it, scientific equations are non-algebraic. There is an educational gap between both communities. At the same time, the mathematical tools cannot work with causal concepts so simple like "symptoms do not cause diseases." This new approach allows Pearl to create algorithms for something so complex and debatable as the transportability of experimental results (Bareinboim and Pearl 2013).

Based in his previously developed structural causal model (SCM), Pearl (2000) combined it with features of the structured equation models (SEM) used in economics and social sciences, as well the potential outcome framework of Neyman/ Rubin with, finally, the graphical models[12] developed for probabilistic reasoning and causal analysis. Although Sewall Wright tried at the beginning of twentieth century to express mathematically causal relationships using a combination of equations and graphs, his methods were not powerful enough. The classic example of the obstacle of causal vocabulary in statistics is the Simpson's Paradox. Besides, even in the case of doing correct causal questions, they are not always able to be answered experimentally. Take, for example, questions of attribution (e.g., what fraction of death cases are *due* to specific exposure?) or susceptibility. Then, to be able to answer we should perform a *probabilistic analysis of counterfactuals*. With a new mathematical language applied to causal notions, and good mathematical machinery, statistics can work with statistical causation. Again, the best solution for a big problem is to combine ideas and innovate methodologically. But as Pearl's mantra affirms: *Think nature, not experiments*. Perhaps is time to admit that all our epistemological tools are provisional and fallible elements and that the path toward better knowledge is necessarily close to a critical thinking. Ontological disambiguation about causality and or statistics will not emanate by itself or due to any

---

[11]Frigg and Hoeffer (2013) talk about several levels of ontological reality that can, consequently, be captured by diverse probability rules, making possible an analysis of chance. Part of their thesis follows the idea that laws and chance rules are formulated in a natural language and that by reformulating Lewis' HBS approach to chance it is possible to accommodate determinism and chance as well as to defend a pluralist but objective way to perform statistical analysis, which they call Theory of Humean Objective Chance (THOC). Pearl tries to solve the natural language problems with his ideas, but is not worried about pluralism, because it is something obvious in scientific practices.

[12]In Pearl (2005), he remarked the usefulness of graphical models, following four reasons: (a) allow modular representation of theories; (b) facilitate the systematic construction of methods, (c) make possible explicit encoding of dependencies, and (d) facilitate efficient inference procedures.

analytical process, instead of it, is an honest and critical activity of plenty of several failures and some successes. Let any one of you who is without priors be the first to throw a formula at the others.

References

Bareinboim, E., & Pearl, J. (2013). A general algorithm for deciding transportability of experimental results. *Journal of causal Inference, 1*(1), 107–134.

Cline, B. L. (1965) *The questioners: Physicists and the quantum theory*, New York: Thomas Y. Crowell Company.

Cousins, R. D. (1995). Why isn't every physicist a Bayesian? *American Journal of Physics, 63*(5), 198–410.

Einstein, A. (1949). Remarks concerning the essays brought together in this cooperative volume. In P. A. Schilpp (ed.), *Albert Einstein: Philosopher-scientist (The library of living philosophers, Vol. VII)*. USA: Open Court.

Funtowicz, S. O., & Ravetz, J. R. (1990). *Uncertainty and quality in science for policy*. The Netherlands: Kluwer Academic Publishers.

Funtowicz, S. O., & Ravetz, J. R. (1994). Uncertainty, complexity and post-normal science. *Environmental Toxicology and Chemistry, 13*(12), 1881–1885.

Horwich, P. (2005). *From a deflationary point of view*. Oxford: OUP.

Howard, D. (2005). Albert Einstein as philosopher of science. *Physics Today, 34*.

Leven, S. J. (1987). Choice and neural process. Unpublished doctoral dissertation, University of Texas at Arlington.

Levine, D. S. (1998) Explorations in common sense and common nonsense, http://www.uta.edu/psychology/faculty/levine/EBOOK/index.htm

Levine, D. S., & Perlovsky, L. I. (2010). Emotion in the pursuit of understanding. *International Journal of Synthetic Emotions, 1*(2), 1–11.

Lovric, M. (2011). *Pro Statistica Scientia*, Pace et Fraternitate Gentium. In *International encyclopedia of statistical science* (p. xvii). Berlin: Springer.

McGrayne, S. B. (2011). *The theory that would not die: How bayes' rule cracked the enigma code, hunted down russian submarines, and emerged triumphant from two centuries of controversy*. USA: Yale University Press.

Pearl, J. (2000). *Causality: Models, reasoning, and inference*. NY: CUP.

Pearl, J. (2001). Bayesianism and causality, or, why i am only a half-bayesian. In D. Corfield & J. Williamson (Eds.), *Foundations of bayesianism* (pp. 12–36). The Netherlands: Kluwer Acamedic Publishers.

Pearl, J. (2005). Influence diagrams-historical and personal perspectives. *Decision Analysis, 2*(4), 232–234.

Pearl, J. (2009). Causal inference in statistics: An overview. *Statistics Surveys, 3*, 96–146.

Poincaré, H. (1914) Science and method, London: T. Nelson and Sons.

Ried, K., et al. (2015). A quantum advantage for inferring causal structure. *Nature Physics, 11*, 414–420.

Scoto, B. (2003). Bayesianism and simplicity. Dissertation: Stanford University.

Suppes, P. (1970). *A probabilistic theory of causality*. Helsinki: North-Holland Publishing Company.

Wagenmakers, E. J., Lee, M. D., Lodewyckx, T., & Iverson, G. (2008). Bayesian versus frequentist inference". In H. Hoijtink, I. Klugkist & P. A. Boeden (Eds.), *Bayesian evaluation of informative hypotheses* (pp. 181–207). NY: Springer.

Wynne, B. (1992). Uncertainty and environmental learning. *Global Environmental Change, 2*, 111–127.

Index